Revise
AS

OCR
Biology

Joh... ...ett

Contents

Chapter 1 Biological molecules

Chapter 2 Cells

Contents

Specification list

The specification labels on each page refer directly to the units in the exam specification, i.e. OCR ▸ 1.2.1 refers to unit 1.2.1.

OCR Biology

MODULE	SPECIFICATION TOPIC	CHAPTER REFERENCE	STUDIED IN CLASS	REVISED	PRACTICE QUESTIONS
Unit 1 Module 1	Cell structure	2.1, 2.2, 2.3			
	Cell membranes	4.1, 4.2			
	Cell division	6.2			
	Cell differentiation	2.3			
Unit 1 Module 2	Gaseous exchange	4.3			
	Transport in animals	5.1, 5.2, 5.3, 5.4			
	Transport in plants	5.5			
Unit 2 Module 1	Biological molecules	1.1, 1.2, 1.3, 1.4, 1.5			
	Nucleic acids	6.1			
	Enzymes	3.1, 3.2			
Unit 2 Module 2	Diet and Food	8.1			
	Disease and immunity	8.1, 8.2, 8.3, 8.4			
Unit 2 Module 3	Biodiversity	7.2			
	Classification	7.1			
	Evolution	7.2			
	Maintaining biodiversity	7.3			

Examination analysis

| Unit 1 | 1 hour written exam
AS Level – 30%
A Level – 15% |
|---|---|
| Unit 2 | 1 hour 45 mins written exam
AS Level – 50%
A Level – 25% |
| Unit 3 | Internal assessment
AS Level – 20%
A Level – 10% |

The AS/A2 Level Biology course

AS and A2

The OCR Biology A Level course being studied from September 2008 is in two parts, with a number of separate modules or units in each part. Most students will start by studying the AS (Advanced Subsidiary) course. Some will go on to study the second part of the A Level course, called A2. Advanced Subsidiary is assessed at the standard expected halfway through an A Level course, i.e. between GCSE and A Level. This means that the new AS and A2 courses are designed so that difficulty steadily increases:

- AS Biology builds from GCSE Science and Additional Science/Biology.
- A2 Biology builds from AS Biology.

How will you be tested?

Assessment units

AS Biology comprises three units or modules. The first two units are assessed by examinations. The third component involves practical assessment in the form of centre-based controlled assessment.

Centre-based controlled assessment involves practical skills marked by your teacher. The marks can be adjusted by moderators appointed by the awarding body.

For AS Biology, you will be tested by three assessment units. For the full A Level in Biology, you will take a further three units. AS Biology forms 50% of the assessment weighting for the full A Level.

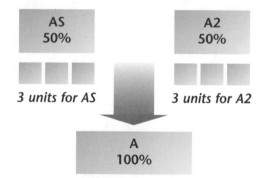

Tests are taken at two specific times of the year, January and June. It can be an advantage to you to take a unit test at the earlier optional time because you can re-sit the test. The best mark from the two will be credited and the lower mark ignored.

Each unit can normally be taken in either January or June. Alternatively, you can study the whole course before taking any of the unit tests. There is a lot of flexibility about when exams can be taken and the following diagram shows just some of the ways that the assessment units may be taken for AS and A Level Biology.

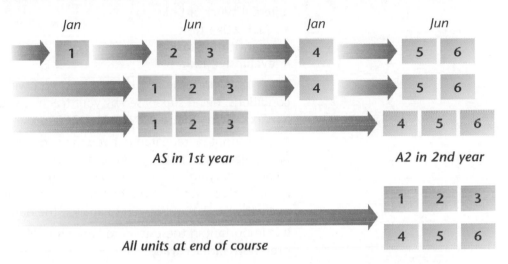

AS in 1st year   *A2 in 2nd year*

All units at end of course

If you are disappointed with a module result, you can re-sit any module. The higher mark counts.

A2 and synoptic assessment

Many students who have studied at AS Level may decide to go on to study A2. There are three further units or modules to be studied. The A Level specification includes a 'synoptic' assessment at the end of A2. Synoptic questions make use of concepts from earlier units, bringing them together in holistic contexts. Examiners will test your ability to inter-relate topics through the complete course from AS to A2.

What skills will I need?

For AS Biology, you will be tested by assessment objectives: these are the skills and abilities that you should have acquired by studying the course. The assessment objectives for AS Biology are shown below.

Knowledge with understanding

- recall of facts, terminology and relationships
- understanding of principles and concepts
- drawing on existing knowledge to show understanding of the responsible use of biological applications in society
- selecting, organising and presenting information clearly and logically

Application of knowledge and understanding, analysis and evaluation

- explaining and interpreting principles and concepts
- interpreting and translating, from one to another, data presented as continuous prose or in tables, diagrams and graphs
- carrying out relevant calculations
- applying knowledge and understanding to familiar and unfamiliar situations
- assessing the validity of biological information, experiments, inferences and statements

You must also present arguments and ideas clearly and logically, using specialist vocabulary where appropriate. Remember to balance your argument!

Experimental and investigative skills

One of the three AS units (and one of the A2 units) is designed to test your experimental and investigative skills. Each unit requires you to carry out three different types of task:
- **qualitative task**
- **quantitative task**
- **evaluative task**

These tasks are provided by OCR and you will carry them out under controlled conditions, being supervised by your teacher. Your teacher then marks them using a mark scheme that is provided by OCR.

A task cannot be repeated but you may be given the opportunity to do more than one of each type of task. Your best mark on each will count. The **qualitative** task involves carrying out a practical task, making and recording observations. The **quantitative** task involves making and recording accurate measurements. The **evaluative** task requires you to analyse the data collected in the quantitative task.

It is important in these tasks to know the difference between **validity**, **reliability**, **precision** and **accuracy**.

Validity: in a valid experiment you only investigate one variable at a time. All the other variables are kept constant. This means that you are really measuring what you are trying to measure.

Reliability: an experiment is reliable if it can be repeated and similar results are obtained. Reliability is usually be increased by taking repeat readings and averaging them.

Accuracy: an accurate measurement is one that is close to an accepted value. This accepted value is usually the work of many scientists and can be checked from accepted sources.

Precision: If an experiment is precise it means that readings can be made to a number of significant digits and that similar results can be made each time.

The best way to describe the difference between accuracy and precision is to use the example of a darts board:

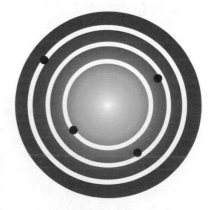

These darts are precise because they are all close to each other, but they are not very accurate.

These darts are precise because they are all close to each other and are accurate.

These darts are not very precise, nor very accurate.

Different types of questions in AS examinations

In AS Biology examinations different types of questions are used to assess your abilities and skills. Unit tests mainly use structured questions requiring both short-answers and more extended answers.

Short-answer questions

A question will normally begin with a brief amount of stimulus material. This may be in the form of a diagram, data or graph. A short-answer question may begin by testing recall. Usually this is followed up by questions which test understanding. Often you will be required to analyse data.

Short-answer questions normally have a space for your responses on the printed paper. The number of lines is a guide as to the amount of words you will need to answer the question. The number of marks indicated on the right side of the paper shows the number of marks you can score for each question part.

Here are some examples. (The answers are shown in *blue*)

The diagram shows part of a DNA molecule.

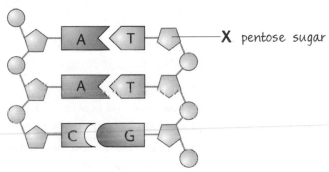

(a) Label part X. [1]

(b) Complete the diagram by writing a letter for each missing organic base in each empty box. [1]

(b) How do two strands of DNA join to each other?
     The organic bases ✓ *link the strands by hydrogen bonds* ✓ [2]

Structured questions

Structured questions are in several parts. The parts are usually about a common context and they often progress in difficulty as you work through each of the parts. They may start with simple recall, then test understanding of a familiar or unfamiliar situation. If the context seems unfamiliar the material will still be centred around concepts and skills from the Biology specification. (If a student can answer questions about unfamiliar situations then they display understanding rather than simple recall.)

When answering structured questions, do not feel that you have to complete a question before starting the next. Answering a part that you are sure of will build your confidence. If you run out of ideas go on to the next question. This will be more profitable than staying with a very difficult question which slows down progress. Return at the end when you have more time.

Here is an example of a structured question which becomes progressively more demanding.

Question

The diagram shows the molecules of a cell surface membrane.

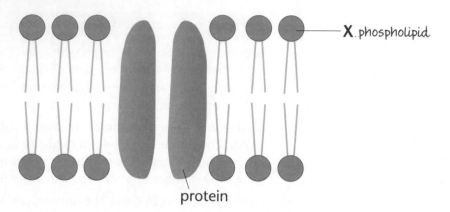

X phospholipid

protein

(a) (i) Label molecule X. [1]

(ii) The part of molecule X facing the outside of a cell is hydrophilic. What does this mean?

water loving/water attracting [1]

(iii) Describe **one** feature of the part of molecule X which faces inwards.

Hydrophobic/ water hating fatty acid residues [1]

(b) Explain how the protein shown in the diagram can actively transport the glucose molecule into the cell.

Energy is released from mitochondria near the channel protein the channel protein opens [3]

Note the help given in diagrams. The labelling of the protein molecule may trigger the memory so that the candidate has to make a small step to link the 'channel' function to this diagram. Examiners give clues! Expect more clues at AS Level than at A2 Level.

Extended answers

In AS Biology, questions requiring more extended answers will usually form part of structured questions. They will normally appear at the end of a structured question and will typically have a value of three to six marks. Longer answers are allocated more lines, so you can use this as a guide as to the extent of your answer. The mark allocation is a guide as to how many points you need to make in your response. Often for an answer worth six marks the mark scheme could have eight creditable answers. You are awarded up to the maximum, six in this instance.

Candidates are assessed on their ability to use a suitable style of writing, and organise relevant material, both logically and clearly. The use of specialist biological words in context is also assessed. Spelling, punctuation and grammar are also taken into consideration. Here is a longer extended response question.

Question

Give an account of the effects of sewage entry into a river and explain the possible consequences to organisms downstream.

The sewage enters the river and is decomposed by bacteria. ✔ *These bacteria are saprobiotic* ✔ *they produce nitrates which act as a fertiliser.* ✔ *Algae form a blanket on the surface* ✔ *light cannot reach plants under the algae so these plants die.* ✔ *Bacteria decompose the dead plants* ✔ *the bacteria use oxygen/bacteria are aerobic* ✔ *fish die due to lack of oxygen* ✔ *Tubifex worms or bloodworms increase near sewage entry* ✔ *mayfly larvae cannot live close to sewage entry/mayfly larvae appear a distance downstream where oxygen levels return.* ✔

10 marking points → [6]

Remember that mark schemes for extended questions often exceed the question total, but you can only be awarded credit up to a maximum. Examiners sometimes build in a hurdle, e.g. in the above responses, references to one organism which increases in population is worthy of a mark, and another which decreases in population is worth another. Continually referring to different species which repeat a growth pattern will not gain further credit.

Exam technique

AS Biology builds from grade C in GCSE Science and GCSE Additional Science (combined) or GCSE Biology. This study guide has been written so that you will be able to tackle AS Biology from a GCSE science background.

You should not need to search for important Biology from GCSE science because this has been included where needed in each chapter. If you have not studied science for some time, you should still be able to learn AS Biology using this text alone.

What are examiners looking for?

Whatever type of question you are answering, it is important to respond in a suitable way. Examiners use instructions to help you to decide the length and depth of your answer. The most common words used are given below, together with a brief description of what each word is asking for.

Define

This requires a formal statement. Some definitions are easy to recall.

> Define the term active transport.
>
> *This is the movement of molecules from where they are in lower concentration to where they are in higher concentration. The process requires energy.*

Other definitions are more complex. Where you have problems it is helpful to give an example.

> Define the term endemic.
>
> *This means that a disease is found regularly in a group of people, district or country. Use of an example clarifies the meaning. Indicating that malaria is invariably found everywhere in a country, confirms understanding.*

Explain

This requires a reason. The amount of detail needed is shown by the number of marks allocated.

> Explain the difference between resolution and magnification.
>
> *Resolution is the ability to be able to distinguish between two points whereas magnification is the number of times an image is bigger than an object itself.*

State

This requires a brief answer without any reason.

> State one role of blood plasma in a mammal.
>
> *Transport of hormones to their target organs.*

List

This requires a sequence of points with no explanation.

> List the abiotic factors which can affect the rate of photosynthesis in pond weed.
>
> *carbon dioxide concentration; amount of light; temperature; pH of water*

Describe

This requires a piece of prose which gives key points. Diagrams should be used where possible.

Describe the nervous control of heart rate.

The medulla oblongata ✓ of the brain connects to the sino atrial node in the right atrium, wall ✓ via the vagus nerve and the sympathetic nerve ✓ the sympathetic nerve speeds up the rate ✓ the vagus nerve slows it down. ✓

Discuss

This requires points both for and against, together with a criticism of each point. (**Compare** is a similar command word.)

Discuss the advantages and disadvantages of using systemic insecticides in agriculture.

Advantages are that the insecticides kill the pests which reduce yield ✓ they enter the sap of the plants so insects which consume sap die ✓ the insecticide lasts longer than a contact insecticide, 2 weeks is not uncommon ✓

Disadvantages are that insecticide may remain in the product and harm a consumer e.g. humans ✓ it may destroy organisms other than the target ✓ no insecticide is 100% effective and develops resistant pests. ✓

Suggest

This means that there is no single correct answer. Often you are given an unfamiliar situation to analyse. The examiners hope for logical deductions from the data given and that, usually, you apply your knowledge of biological concepts and principles.

The graph shows that the population of lynx decreased in 1980. Suggest reasons for this.

Weather conditions prevented plant growth ✓ so the snowshoe hares could not get enough food and their population remained low ✓ so the lynx did not have enough hares(prey) to predate upon. ✓ The lynx could have had a disease which reduced numbers. ✓

Calculate

This requires that you work out a numerical answer. Remember to give the units and to show your working, marks are usually available for a partially correct answer. If you work everything out in stages write down the sequence. Otherwise if you merely give the answer and it is wrong, then the working marks are not available to you.

Calculate the Rf value of spot X. (X is 25 mm from start and solvent front is 100 mm)

$$Rf = \frac{\text{distance moved by spot}}{\text{distance moved by the solvent front}}$$

$$= \frac{25 \text{ mm}}{100 \text{ mm}}$$

$$= 0.25$$

Outline

This requires that you give only the main points. The marks allocated will guide you on the number of points which you need to make.

Outline the use of restriction endonuclease in genetic engineering.

The enzyme is used to cut the DNA of the donor cell. ✓

It cuts the DNA up like this A T | G C C G A T   =   A T   + G C C G A T ✓
                                    T A C G G C | T A   T A C G G C   T A

The DNA in a bacterial plasmid is cut with the same restriction endonuclease. ✓

The donor DNA will fit onto the sticky ends of the broken plasmid. ✓

If a question does not seem to make sense, you may have mis-read it. Read it again!

Some dos and don'ts

Dos

Do answer the question

No credit can be given for good Biology that is irrelevant to the question.

Do use the mark allocation to guide how much you write

Two marks are awarded for two valid points – writing more will rarely gain more credit and could mean wasted time or even contradicting earlier valid points.

Do use diagrams, equations and tables in your responses

Even in 'essay style' questions, these offer an excellent way of communicating biology.

Do write legibly

An examiner cannot give marks if the answer cannot be read.

Do write using correct spelling and grammar. Structure longer essays carefully

Marks are now awarded for the quality of your language in exams.

Don'ts

Don't fill up any blank space on a paper

In structured questions, the number of dotted lines should guide the length of your answer.

If you write too much, you waste time and may not finish the exam paper. You also risk contradicting yourself.

Don't write out the question again

This wastes time. The marks are for the answer!

Don't contradict yourself

The examiner cannot be expected to choose which answer is intended. You could lose a hard-earned mark.

Don't spend too much time on a part that you find difficult

You may not have enough time to complete the exam. You can always return to a difficult calculation if you have time at the end of the exam.

What grade do you want?

Everyone would like to improve their grades but you will only manage this with a lot of hard work and determination. You should have a fair idea of your natural ability and likely grade in biology and the hints below offer advice on improving that grade.

For a Grade A

You will need to be a very good all-rounder.

- You must go into every exam knowing the work extremely well.
- You must be able to apply your knowledge to new, unfamiliar situations.
- You need to have practised many, many exam questions so that you are ready for the type of question that will appear.

The exams test all areas of the syllabus and any weaknesses in your biology will be found out. There must be no holes in your knowledge and understanding. For a Grade A, you must be competent in all areas.

For a Grade C

You must have a reasonable grasp of biology but you may have weaknesses in several areas and you will be unsure of some of the reasons for the biology.

- Many Grade C candidates are just as good at answering questions as the Grade A students but holes and weaknesses often show up in just some topics.
- To improve, you will need to master your weaknesses and you must prepare thoroughly for the exam. You must become a better all-rounder.

For a Grade E

You cannot afford to miss the easy marks. Even if you find biology difficult to understand and would be happy with a Grade E, there are plenty of questions in which you can gain marks.

- You must memorise all definitions.
- You must practise exam questions to give yourself confidence that you do know some biology. In exams, answer the parts of questions that you know first. You must not waste time on the difficult parts. You can always go back to these later.
- The areas of biology that you find most difficult are going to be hard to score on in exams. Even in the difficult questions, there are still marks to be gained. Show your working in calculations because credit is given for a sound method. You can always gain some marks if you get part of the way towards the solution.

What marks do you need?

The table below shows how your average mark is transferred into a grade.

average	80%	70%	60%	50%	40%
grade	A	B	C	D	E

To achieve an A* grade, you need to achieve a...

- grade A overall (80% or more on uniform mark scale) for the **whole** A level qualification
- grade A* (90% or more on the uniform mark scale) across your A2 units.

A* grades are awarded for the A level qualification only and not for the AS qualification or individual units.

Four steps to successful revision

Step 1: Understand

- Study the topic to be learned slowly. Make sure you understand the logic or important concepts.
- Mark up the text if necessary – underline, highlight and make notes.
- Re-read each paragraph slowly.

GO TO STEP 2

Step 2: Summarise

- Now make your own revision note summary:
  What is the main idea, theme or concept to be learnt?
  What are the main points? How does the logic develop?
  Ask questions: Why? How? What next?
- Use bullet points, mind maps, patterned notes.
- Link ideas with mnemonics, mind maps, crazy stories.
- Note the title and date of the revision notes
  (e.g. Biology: Cells, 3rd March).
- Organise your notes carefully and keep them in a file.

This is now in **short-term memory**. You will forget 80% of it if you do not go to Step 3.
GO TO STEP 3, but first take a 10 minute break.

Step 3: Memorise

- Take 25 minute learning 'bites' with 5 minute breaks.
- After each 5 minute break test yourself:
  Cover the original revision note summary.
  Write down the main points.
  Speak out loud (record on tape).
  Tell someone else.
  Repeat many times.

The material is well on its way to **long-term memory**.
You will forget 40% if you do not do step 4. **GO TO STEP 4**

Step 4: Track/Review

- Create a Revision Diary (one A4 page per day).
- Make a revision plan for the topic, e.g. 1 day later, 1 week later, 1 month later.
- Record your revision in your Revision Diary, e.g.
  Biology: Cells, 3rd March 25 minutes
  Biology: Cells, 5th March 15 minutes
  Biology: Cells, 3rd April 15 minutes
  ... and then at monthly intervals.

Chapter 1
Biological molecules

The following topics are covered in this chapter:

- Carbohydrates
- Lipids
- Proteins

- The importance of water to life
- Biochemical tests

1.1 Carbohydrates

After studying this section you should be able to:

- recall the main elements found in carbohydrates
- recall the structure of glucose, fructose, lactose, sucrose, starch, glycogen and cellulose
- recall the role of glucose, starch, glycogen, cellulose and pectin

LEARNING SUMMARY

Structure of carbohydrates

OCR   2.1.1

Monosaccharides

All carbohydrates are formed from the elements carbon (C), hydrogen (H) and oxygen (O). The formula of a carbohydrate is always $(CH_2O)_n$. The n represents the number of times the basic CH_2O unit is repeated, e.g. where n = 6 the molecular formula is $C_6H_{12}O_6$. This is the formula shared by glucose and other simple sugars like fructose. These simple sugars are made up from a single sugar unit and are known as **monosaccharides**.

The molecular formula, $C_6H_{12}O_6$, does not indicate how the atoms bond together. Bonded to the carbon atoms are a number of $-H$ and $-OH$ groups. Different positions of these groups on the carbon chain are responsible for different properties of the molecules. The structural formulae of α and β glucose are shown below.

These molecules are mirror images of each other. When molecules have the same molecular formula but different structural formulae, they are known as **isomers**. Isomers have different properties to each other.

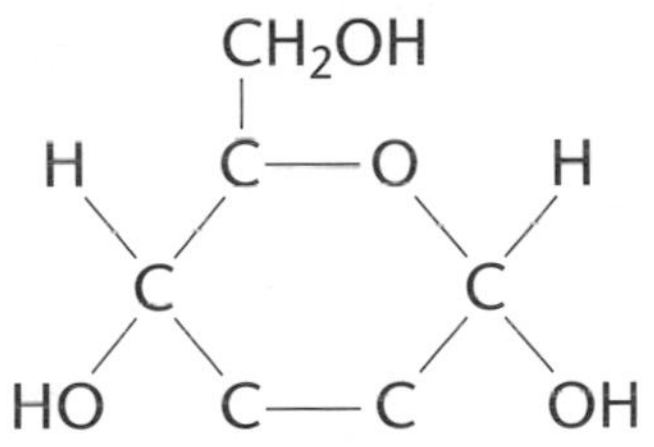

α glucose

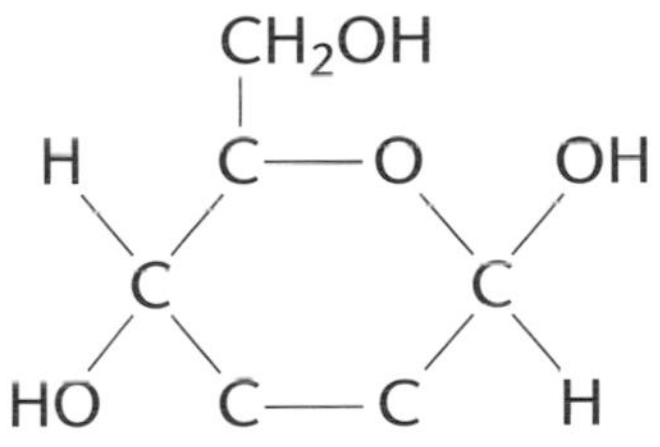

β glucose

Glucose is so small that it can pass through the villi and capillaries into our bloodstream. The molecules subsequently release energy as a result of respiration. Simple glucose molecules are capable of so much more than just releasing energy. They can combine with others to form bigger molecules.

17

Disaccharides

Each glucose unit is known as a **monomer** and is capable of linking others. This diagram shows two molecules of α glucose forming a **disaccharide**.

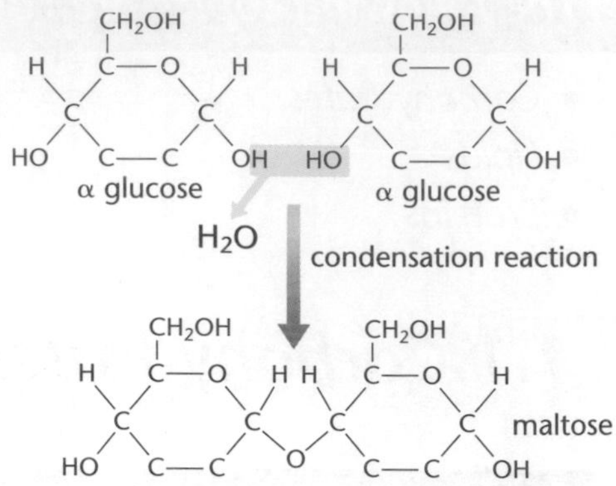

In your examinations look for different monosaccharides being given, like fructose or β glucose. You may be asked to show how they bond together. The principle will be exactly the same.

A **condensation** reaction means that as two carbohydrate molecules bond together a water molecule is produced. The link formed between the two glucose molecules is known as a **glycosidic bond**.

A glycosidic bond can also be broken down to release separate monomer units. This is the opposite of the reaction shown above. Instead of water being given off, a water molecule is needed to break each glycosidic bond. This is called **hydrolysis** because water is needed to split up the bigger molecule.

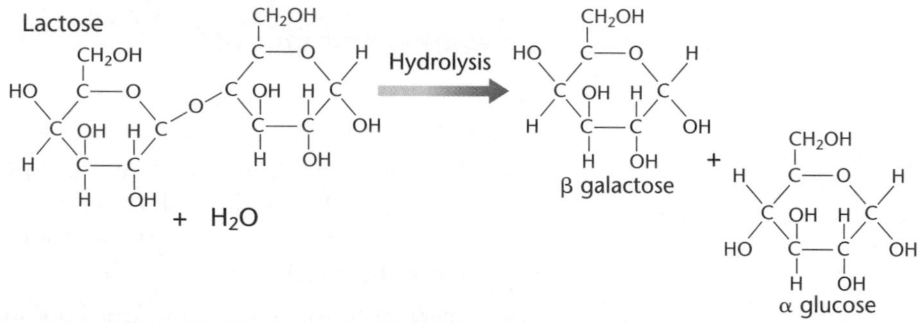

'Lysis' literally means **'splitting'**. In hydrolysis water is needed in the reaction to break down the molecule.

Different disaccharides are made by joining together different monosaccharides.

Disaccharide	Component monosaccharides
lactose	glucose + galactose
maltose	glucose + glucose
sucrose	glucose + fructose

Polysaccharides

Like disaccharides, **polysaccharides** consist of monomer units linked by the glycosidic bond. However, instead of just two monomer units they can have many. Chains of these 'sugar' units are known as **polymers**. These larger molecules have important structural and storage roles.

Starch is a polymer of the sugar, α glucose. The diagram below shows part of a starch molecule.

Notice the five glycosidic bonds on just a small part of a starch molecule.

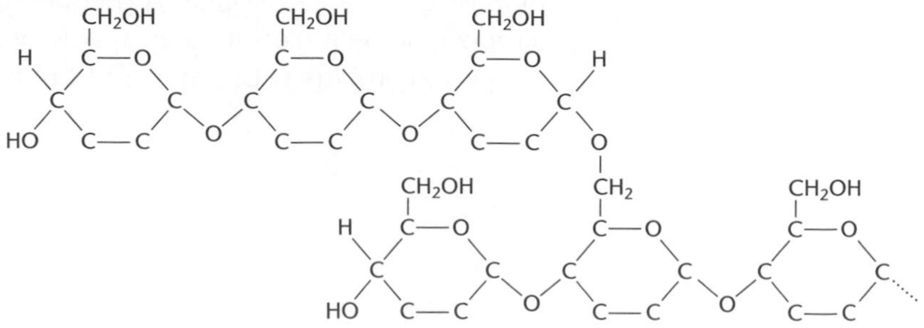

part of a branched section of a starch molecule

This type of starch molecule is called **amylopectin** and it has a branched structure. Starch also contains **amylose**. This does not contain branches but the chain of glucose units forms a helix. **Glycogen** is similar in structure to amylopectin but with more branches. **Cellulose** is also a polymer of glucose units, but this time the units are β glucose.

How useful are polysaccharides?

- **Starch** is stored in organisms as a future energy source, e.g. potato has a high starch content to supply energy for the buds to grow at a later stage.
- **Glycogen** is stored in the liver, which releases glucose for energy in times of low blood sugar.

Both starch and glycogen are insoluble which enables them to remain inside cells.

The many branches in the amylopectin molecule means that enzymes can digest the molecule rapidly.

- **Cellulose** has long molecules which help form a tough protective layer around plant cells, the cell wall. Each cellulose molecule has up to 10 000 β glucose units. Each molecule can form cross-links with other cellulose molecules forming fibres. This makes cellulose fibres very strong.

Pectins help cells to bind together.

- **Pectins** are used alongside cellulose in the cell wall. They are polysaccharides which are bound together by calcium pectate.

Together the cellulose and pectins give exceptional mechanical strength. The cell wall is also permeable to a wide range of substances.

1.2 Lipids

After studying this section you should be able to:

LEARNING SUMMARY

- *recall the main elements found in lipids*
- *recall the structure of lipids*
- *distinguish between saturated and unsaturated fats*
- *recall the role of lipids*

What are lipids?

OCR 2.1.1

Lipids include **oils**, **fats** and **waxes**. They consist of exactly the same elements as carbohydrates, i.e. carbon (C), hydrogen (H) and oxygen (O), but their proportion is different. There is always a high proportion of carbon and hydrogen, with a small proportion of oxygen.

The diagram below shows the structural formula of the most common type of lipid called a **triglyceride**.

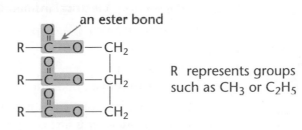

R represents groups such as CH_3 or C_2H_5

a **triglyceride** fat

Triglycerides are formed when fatty acids react with glycerol. During this reaction water is produced, a further example of a condensation reaction. The essential bond is the **ester bond**.

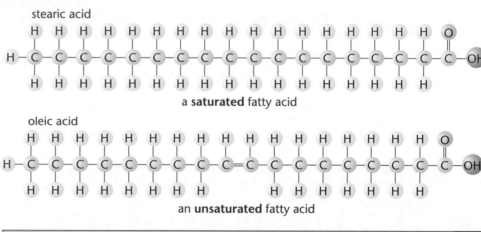

3 fatty acids   glycerol   a triglyceride fat   water

> Note that water is produced during triglyceride formation. This is another example of a condensation reaction. Different triglyceride fats are formed from different fatty acids.

Triglycerides can be changed back into the original fatty acids and glycerol. Enzymes are needed for this transformation together with water molecules. Remember, an enzyme reaction which requires water to break up a molecule is known as **hydrolysis**.

What are saturated and unsaturated fats?

The answer lies in the types of fatty acid used to produce them.

> The hydrocarbon chains are so long that they are often represented by the acid group (–COOH) and a zig-zag line.
>
> unsaturated
> ∧∧∧∧=∧∧∧COOH
>
> saturated
> ∧∧∧∧∧∧∧COOH

stearic acid

a **saturated** fatty acid

oleic acid

an **unsaturated** fatty acid

> Saturated fats lead to cholesterol build up in the blood vessels so many people try to avoid them in their diet. Blocked vessels cause heart attacks.

> Note that when saturated fatty acids react with glycerol they make saturated fats. Conversely, unsaturated fatty acids make unsaturated fats.

KEY POINT

Saturated fatty acids have no C=C (double bonds) in their hydrocarbon chain, but unsaturated fatty acids do. This is the difference.

How useful are lipids?

Like carbohydrates, lipids are used as an **energy** supply, but a given amount of lipid releases more energy than the same amount of carbohydrate. Due to their **insolubility** in water and **compact** structure, lipids have long-term **storage** qualities. Adipose cells beneath our skin contain large quantities of fat which **insulate** us and help to maintain body temperature. Fat gives **mechanical** support around our soft organs and even gives **electrical insulation** around our nerve axons.

An aquatic organism such as a dolphin has a large fat layer which:

- is an energy store
- is a thermal insulator
- helps the animal remain buoyant.

The most important role of lipids is their function in cell membranes. To fulfil these functions a triglyceride fat is first converted into a **phospholipid**.

> Never give an ambiguous answer like this: 'fats keep an animal warm'. The examiner does not know whether you are referring to energy release by mitochondria or the insulation property. Be specific, otherwise you will not be credited.

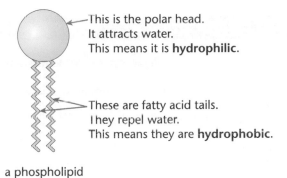

The chemical reaction diagram shows:

$$H_2C-C(H)-CH_2$$ with O, O, O and $C=O$ $C=O$ $C=O$ (triglyceride) $+$ $O=P-OH$ with OH, OH (phosphoric acid) $\rightarrow$ $H_2C-C(H)-CH_2$ with O, O, O and $O=P-O$ $C=O$ $C-O$, O (phospholipid)

triglyceride   phosphoric acid   phospholipid

Phosphoric acid replaces one of the fatty acids of the triglyceride. The new molecule, the phospholipid, is a major component of cell membranes. Cell membranes also contain the lipid cholesterol. The diagram below represents a phospholipid.

This is the polar head.
It attracts water.
This means it is **hydrophilic**.

These are fatty acid tails.
They repel water.
This means they are **hydrophobic**.

a phospholipid

> Remember this simplified diagram when you learn about the cell membrane.

1.3 Proteins

After studying this section you should be able to:

- recall the main elements found in proteins
- recall how proteins are constructed
- recall the structure of proteins
- recall the major functions of proteins

LEARNING SUMMARY

The building blocks of proteins

OCR   2.1.1

Like carbohydrates and lipids, proteins contain the elements carbon (C), hydrogen (H) and oxygen (O), but in addition they **always** contain **nitrogen** (N). Sulfur is also often present.

Before understanding how proteins are constructed, the structure of **amino acids** should be noted. The diagram below shows the general structure of an amino acid.

> Just like the earlier carbohydrate and lipid molecules, 'R' represents groups such as –CH₃ and –C₂H₅. There are about 20 commonly found amino acids but you will not need to know them all. Instead, learn the basic structure shown opposite.

amine group   carboxylic acid group

$$H-N(H)-C(R)(H)-C(=O)-OH$$

an amino acid

How is a protein constructed?

The process begins by amino acids bonding together. The diagram shows two amino acids being joined together by a **peptide bond**.

This is another example of a condensation reaction as water is produced as the dipeptide molecule is assembled.

Note that the peptide bonds can be broken down by a hydrolysis reaction.

amino acid   amino acid   a dipeptide   + H₂O

When many amino acids join together a long-chain **polypeptide** is produced. The linking of amino acids in this way takes place during protein synthesis (see page 73). There are around 20 different amino acids. Organisms join amino acids in different linear sequences to form a variety of polypeptides, then build these polypeptides into complex molecules, the **proteins**. Humans need eight essential amino acids as adults and ten as children, all the others can be made inside the cells.

The sequence of amino acids along a polypeptide is controlled by another complex molecule, DNA (see the genetic code, page 72).

Levels of organisation in proteins

OCR   2.1.1

Primary protein structure

This is the **linear sequence** of amino acids.

peptide bond   amino acid

primary structure

Secondary protein structure

Polypeptides become twisted or coiled. These shapes are known as the **secondary structure**. There are two common secondary structures; the α-helix and the β-pleated sheet.

Both secondary structures give additional strength to proteins. The α-helix helps make tough fibres like the protein in your nails, e.g. keratin. The β-pleated sheet helps make the strength-giving protein in silk, fibroin. Many proteins are made from both α-helix and β-pleated sheet.

amino acid

hydrogen bonds hold shape together

α-**helix**

amino acid

hydrogen bonds hold shape together

β-**pleated sheet**

The polypeptides are held in position by hydrogen bonds. In both α-helices and β-pleated sheets the **C = O** of one amino acid bonds to the **H–N** of an adjacent amino acid like this: C = O --- H–N.

----- = hydrogen bonds

An α-helix is a tight, twisted strand; a β-pleated sheet is where a zig-zag line of amino acids bonds with the next, and so on. This forms a sheet or ribbon shape.

The protein shown only achieves a secondary structure as the simple α-helix polypeptides do not undergo further folding.

This is the structure of collagen, a fibrous protein. It is made of three α-helix polypeptides twisted together.

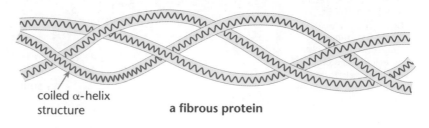

coiled α-helix structure   **a fibrous protein**

Note that the specific contours of proteins have extremely significant roles in life processes. (See enzymes page 40.)

Tertiary protein structure

This is when a polypeptide is **folded** into a **precise** shape. The polypeptide is held in 'bends' and 'tucks' in a **permanent** shape by a range of bonds including:

- **disulfide** bridges (sulfur–sulfur bonds)
- hydrogen bonds
- ionic bonds
- hydrophobic and hydrophilic interactions.

Some proteins are folded up into a spherical shape. They are called **globular proteins**. They are soluble in water. Other proteins, called **fibrous proteins**, form long chains. They are insoluble and usually perform structural functions.

This is the structure of a **globular** protein. It is made of an α-helix and a β-pleated sheet. Precise shapes are formed with specific contours.

Quaternary protein structure

Some proteins consist of **different polypeptides** bonded together to form extremely intricate shapes. A haemoglobin molecule is formed from four separate polypeptide chains. It also has a haem group, which contains iron. This inorganic group is known as a **prosthetic group** and in this instance aids oxygen transport.

Note that some proteins do not have a quaternary structure. If they consist of just one folded polypeptide then they are classified as having tertiary structure. If they are simple fibres of α-helices or β-pleated sheets then they have only secondary protein structure.

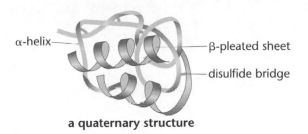

α-helix — β-pleated sheet — disulfide bridge

a quaternary structure

How useful are proteins?

OCR 2.1.1

Proteins can be just as beneficial as carbohydrates and lipids in releasing energy. Broken down into their component amino acids, they liberate energy during respiration. The list below shows **important** uses of proteins:

- **cell-membrane proteins** transport substances across the membrane for processes such as facilitated diffusion and active transport
- **enzymes** catalyse biochemical reactions, e.g. pepsin breaks down protein into polypeptides
- **hormones** are passed through the blood and trigger reactions in other parts of the body, e.g. insulin regulates blood sugar
- **immuno-proteins**, e.g. antibodies are made by lymphocytes and act against antigenic sites on micro-organisms
- **structural proteins** give strength to organs, e.g. collagen makes tendons tough
- **transport proteins**, e.g. haemoglobin transports oxygen in the blood
- **contractile proteins**, e.g. actin and myosin help muscles shorten during contraction
- **storage proteins**, e.g. aleurone in seeds helps germination, and casein in milk helps supply valuable protein to babies
- **buffer proteins**, e.g. blood proteins, due to their charge, help maintain the pH of plasma.

Progress check

1 List the sequence of structures in a globular protein such as haemoglobin.

2 The following statements refer to proteins used for different functions in the body. The list gives the name of different types of protein. Match the name of each type of protein with the correct statement.

(i)   transport proteins
(ii) immuno-proteins
(iii) storage proteins
(iv) buffer proteins
(v)   cell-membrane proteins

(vi) contractile proteins
(vii) enzymes
(viii) structural proteins
(ix) hormones

A haemoglobin is used to transport oxygen in blood.

B aleurone in seeds is a source of amino acids as it is broken down during germination.

C actin and myosin help muscles shorten during contraction.

D antibodies made by lymphocytes against antigens.

E blood proteins, due to their charge, help maintain the pH of plasma.

F used to transport substances across the membrane for processes such as facilitated diffusion.

G passed through blood, used to trigger reactions in other parts of the body, e.g. FSH stimulates a primary follicle.

H used to catalyse biochemical reactions, e.g. amylase breaks down starch into maltose.

I used to give strength to organs, e.g. collagen makes tendons tough.

1.4 The importance of water to life

After studying this section you should be able to:

- *recall the properties of water*
- *recall the functions of water*

LEARNING SUMMARY

Properties and uses of water

OCR ▶ 2.1.1

Water is essential to living organisms. The list below shows some of its properties and uses.

- **Hydrogen bonds** are formed between the oxygen of one water molecule and the hydrogen of another. As a result of this water molecules have an attraction for each other known as **cohesion**.

- **Cohesion** is responsible for surface tension which enables aquatic insects like pond skaters to walk on a pond surface. It also aids capillarity, the way in which water moves through xylem in plants.

- Water is a **dipolar** molecule, which means that the oxygen has a slight negative charge at one end of the molecule, and each hydrogen has a slight positive charge at the other end.

> Try to learn all of the functions of water molecules given in the list. Water is used in so many ways that the chance of being questioned on the topic is high.

- Other **polar** molecules dissolve in water. The different charges on these molecules enable them to fit into water's hydrogen bond structure. Ions in solution can be transported or can take part in reactions. Polar substances can dissolve in water and are called **hydrophilic**. Non-polar substances cannot dissolve in water and are **hydrophobic**.
- Water is used in **photosynthesis**, so it is necessary for the production of glucose. This in turn is used in the synthesis of many chemicals.
- Water helps in the **temperature regulation** of many organisms. It enables the cooling down of some organisms. Owing to a **high latent heat of vaporisation**, large amounts of body heat are needed to evaporate a small quantity of water. Organisms like humans cool down effectively but lose only a small amount of water in doing so.
- A relatively high level of heat is needed to raise the temperature of water by a small amount due to its **high specific heat capacity**. This enables organisms to control their body temperature more effectively.
- Water is a solvent for ionic compounds. A number of the essential elements required by organisms are obtained in ionic form, e.g.:
  (a) plants absorb nitrate ions (NO_3^-) and phosphate ions (PO_4^-) in solution
  (b) animals intake sodium ions (Na^+) and chloride ions (Cl^-).

1.5 Biochemical tests

After studying this section you should be able to:

- describe biochemical tests for carbohydrates, lipids and proteins

LEARNING SUMMARY

Biochemical tests

OCR 2.1.1

Tests for carbohydrates in the laboratory

Benedict's test used to identify reducing sugars (monosaccharides and some disaccharides)
- Add Benedict's solution to the chemical sample and heat.
- The solution changes from blue to brick-red or yellow if a reducing sugar is present.

Non-reducing sugar test used to test for non-reducing sugars, e.g. the disaccharide, sucrose
- First a Benedict's test is performed.
- If the Benedict's test is negative, the sample is hydrolysed by heating with hydrochloric acid, then neutralised with sodium hydrogen carbonate.
- This breaks the glycosidic bond of the disaccharide, releasing the monomers.
- A second Benedict's test is performed which will be positive because the monomers are now free.

Starch test
- Add iodine solution to the sample.
- If starch is present the colour changes to blue-black.

> All the biochemical tests need to be learned. This work is good value because they are regularly tested in 2 or 3 mark question components.

Tests for lipids in the laboratory

Emulsion test used to identify fats and oils
- Add ethanol to the sample, shake, then pour the mixture into water.
- If fats or oils are present then a white emulsion appears at the surface.

Tests for proteins in the laboratory

Biuret test used to identify any protein
- Add dilute sodium hydroxide and dilute copper sulfate to the sample.
- A violet colour appears if a protein is present.

Quantitative tests

The above tests can be used to see if the different chemicals are present or absent. They are called **qualitative tests**.

They do not provide any information about how much of the chemical is present unless they are modified. One way to do this is to measure the intensity of the colour produced. This can be done with a machine called a **colorimeter**. It shines a light through the solution and measures how much light gets through. The lighter the colour, the more light passes through and so the less chemical is present. This is a **quantitative test**.

Sample questions and model answers

1 Below are the structures of two glucose molecules.

(a) Complete the equation to show how the molecules react to form a glycosidic bond and the molecule produced.

Remember that you will be given molecule structures. These stimulate your memory which helps you work out the answer.

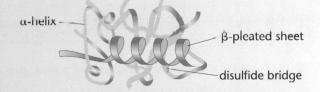

maltose + H_2O

(b) Which form of glucose molecules is shown?
Give a reason for your answer. [2]
α glucose, because the −OH groups on carbon atom 1 are down

The correct answer here is condensation. A regular error in questions like this is to give the wrong reaction, i.e. hydrolysis. Revise carefully then you will make the correct choice.

(c) State the type of reaction which takes place when the two molecules shown above react together. [2]
Condensation.

2 The diagram below shows a globular protein consisting of four polypeptide chains.

α-helix –

β-pleated sheet

disulfide bridge

Look out for similar structures in your examinations. The proteins given may be different, but the principles remain the same.

(a) Use your own knowledge and the information given to explain how this protein shows primary, secondary, tertiary and quaternary structure. [5]
Primary structure: it is formed from chains of amino acids; it has polypeptides made of a linear sequence of amino acids.
Secondary structure: it has an α-helix, it has a β pleated sheet.
Tertiary structure: the polypeptides are folded, the folds are held in position by disulfide bridges.
Quaternary structure: there are four polypeptides in this protein.
Two or more are bonded together to give a quaternary structure.

(b) Name and describe a test which would show that haemoglobin is a protein. [3]
The Biuret test.
Take a sample of haemoglobin and add water, sodium hydroxide and copper sulfate.
The colour of the mixture shows as violet or mauve if the sample is a protein.

Most examinations include at least one biochemical test.

Practice examination questions

Try all of the questions and check your answers with the mark scheme on page 106.

1 (a) Complete the equation below to show the breakdown of a triglyceride fat into fatty acids and glycerol. [2]

a triglyceride fat   3 molecules
of water

(b) Describe a biochemical test which would show if a sample contained fat. [3]

2 The diagram below shows a polypeptide consisting of 15 amino acids.

amino acid

(group X)

(a) Name the bond between each pair of amino acids in this polypeptide. [1]

(b) What is group X? [1]

(c) Which level of protein structure is shown by this polypeptide? Give a reason for your answer. [2]

3 Explain how the following properties of water are useful to living organisms:

(a) a large latent heat of evaporation [2]

(b) a high specific heat capacity [2]

(c) the cohesive attraction of water molecules for each other. [2]

Chapter 2

Cells

The following topics are covered in this chapter:

- *The ultra-structure of cells*
- *Observing cells*

- *Specialisation of cells*

2.1 The ultra-structure of cells

After studying this section you should be able to:

- *identify cell organelles and understand their roles*

LEARNING SUMMARY

Cell organelles

OCR   1.1.1

The cell is the basic functioning unit of organisms in which chemical reactions take place. These reactions involve energy release needed to support life and build structures. Organisms consist of one or more cells. The amoeba is composed of one cell, whereas millions of cells make up a human.

> Every cell possesses internal coded instructions to control cell activities and development. Cells also have the ability to continue life by some form of cell division.
>
> KEY POINT

Organelles are best seen with the aid of an electron microscope.

The ultra-structure of a cell can be seen using an electron microscope. Sub-cellular units called **organelles** become visible. Each organelle has been researched to help us understand more about the processes of life.

The animal cell and its organelles

The diagram below shows the organelles found in a typical animal cell.

A plant cell has all of the same structures plus:
- *a cellulose cell wall*
- *chloroplasts (some cells)*
- *a sap vacuole with tonoplast.*

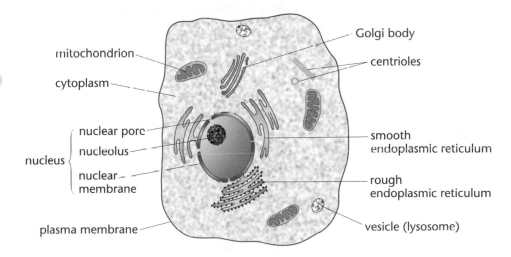

mitochondrion

cytoplasm

Golgi body

centrioles

nucleus {
nuclear pore
nucleolus
nuclear membrane

smooth endoplasmic reticulum

rough endoplasmic reticulum

plasma membrane

vesicle (lysosome)

Cell surface (plasma) membrane

This covers the outside of a cell and consists of a **double** layered sheet of phospholipid molecules interspersed with proteins. It separates the cell from the outside environment, gives physical protection and allows the import and export of **selected** chemicals.

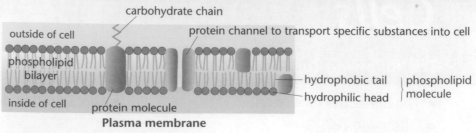

Plasma membrane

Nucleus

This controls all cellular activity using coded instructions located in DNA. These coded instructions enable the cell to make specific proteins. RNA is produced in the nucleus and leaves via the nuclear pores. The nucleus stores, replicates and decodes DNA.

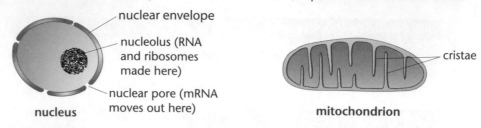

Be ready to identify all cell organelles in either a diagram or electron micrograph. A mitochondrion is often sausage shaped but the end view is circular. Look out for the internal membranes.

Mitochondria

These consist of a double membrane enclosing a semi-fluid matrix. Throughout the matrix is an internal membrane, folded into cristae. The cristae and matrix contain enzymes which enable this organelle to carry out aerobic respiration. It is the key organelle in the release of energy, making ATP available to the cell.

Mitochondria are needed for many energy requiring processes in the cell, including active transport and the movement of cilia.

Cytoplasm

Cytoplasm is often seen as grey and granular. If the image is 'clear' then you are probably looking at a vacuole.

Each organelle in a cell is suspended in a semi-liquid medium, the cytoplasm. Many ions are dissolved in it. It is the site of many chemical reactions.

Ribosomes

Look for tiny dots in the cytoplasm. They will almost certainly be ribosomes. A membrane adjacent to a line of ribosomes is probably the rough endoplasmic reticulum.

There are numerous **ribosomes** in a cell, located along **rough endoplasmic reticulum**. They aid the manufacture of proteins, being the site where mRNA meets tRNA so that amino acids are bonded together.

Endoplasmic reticulum (ER)

This is found as **rough ER** (with ribosomes) and **smooth ER** (without ribosomes). It is a series of folded internal membranes. Substances are transported in the spaces between the ER. The smooth ER aids the synthesis and transport of lipids.

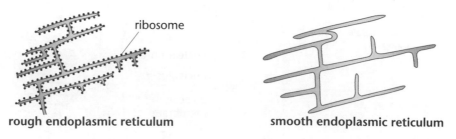

Golgi body

Look for vesicles 'pinching off' the main Golgi sacs.

This is a series of flattened sacs, each separated from the cytoplasm by a membrane. The **Golgi body** is a packaging system where important chemicals become membrane wrapped, forming **vesicles**. The vesicles become detached from the main Golgi sacs, enabling the isolation of chemicals from each other in the cytoplasm. The Golgi body aids the production and secretion of many proteins, carbohydrates and glycoproteins. Vesicle membranes merge with the plasma membrane to enable secretions to take place.

Lysosomes are specialised vesicles because they contain digestive enzymes.

lysosome

Golgi body

Centrioles, cilia and flagella

Several structures in the cell are made of **microtubules** which are made of a protein called tubulin.

Centrioles are found in animal cells and are two short cylinders located near the nucleus. Their function is to organise the spindle which is used in cell division.

Cilia and **flagella** are long, thin extensions from the cell, which can produce movement. They are both made of microtubules but flagella are longer.

There are also microtubules in the cytoplasm. Along with thinner proteins called microfilaments they make up the **cytoskeleton** of the cell. They can move organelles around in the cell.

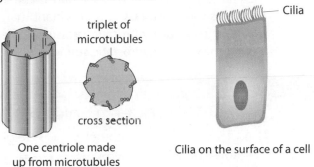

One centriole made up from microtubules

Cilia on the surface of a cell

Section through a cilia showing microtubules

The plant cell and its organelles

All of the structures described for animal cells are also found in plant cells (except the centrioles). Additionally there are three extra structures shown in the diagram below.

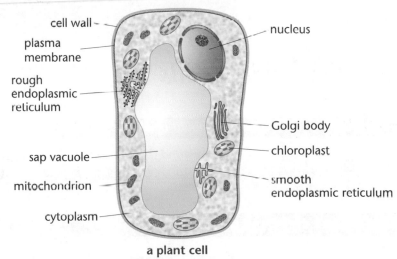

a plant cell

> Did you spot the three extra structures in the plant cell? Remember that a root cell under the soil will not possess chloroplasts. Nor does every plant cell above the soil have chloroplasts.

Cell wall

> Cells other than plant cells can have cell walls, e.g. bacteria have polysaccharides other than cellulose.

Around the plasma membrane of plant cells is the cell wall. This is secreted by the cell and consists of cellulose microfibrils embedded in a layer of calcium pectate and hemicelluloses. Between the walls of neighbouring cells calcium pectate cements one cell to the next in multi-cellular plants. Plant cell walls provide a rigid support for the cell but allow many substances to be imported or exported by the cell. The wall allows the cell to build up an effective hydrostatic skeleton. Some plant cells have a cytoplasmic link which crosses the wall. These links of cytoplasm are known as **plasmodesmata**.

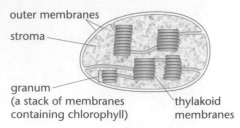

middle lamella

plasmodesmata
(a strand of cytoplasm
connected to next cell)

plasmodesmata

outer membranes

stroma

granum
(a stack of membranes
containing chlorophyll)

thylakoid
membranes

a chloroplast

Chloroplasts

These enable the plant to photosynthesise, making glucose. Each consists of an outer covering of two membranes. Inside are more membranes stacked in piles called **grana**. The membranes enclose a substance vital to photosynthesis, **chlorophyll**. Inside the chloroplast is a matrix known as the **stroma** which is also involved in photosynthesis.

Sap vacuole

This is a large space in a plant cell, containing chemicals such as glucose and mineral ions in water. This solution is the sap. It is surrounded by a membrane known as the **tonoplast**. It is important that a plant cell contains enough water to maintain internal hydrostatic pressure. When this is achieved the cell is turgid, having maximum hydrostatic strength.

Progress check

1 Describe the function of each of the following cell organelles:

nucleus   centrioles   Golgi body
mitochondria   ribosomes   cell (plasma) membrane

2 Give **three** structural differences between a plant and animal cell.

2 A plant cell has a cellulose cell wall, chloroplasts, and a sap vacuole lined by a tonoplast.
Cell (plasma) membrane – gives physical protection to the outside of a cell, allows the import and export of *selected chemicals*.
Golgi body – is a packaging system where chemicals become membrane wrapped, forming vesicles
ribosomes – aid the manufacture of proteins, being the site where mRNA meets tRNA so that amino acids are bonded together
centrioles – help produce the spindle during cell division
mitochondria – release energy during aerobic respiration
1 **nucleus** – mRNA is produced in the nucleus with the help of DNA

2.2 Observing cells

After studying this section you should be able to:

- *understand the principles of light and electron microscopes*
- *understand how microscopic specimens are measured*

LEARNING SUMMARY

Light and electron microscopy

OCR ▶ 1.1.1

Microscopes **magnify** the image of a specimen to enable the human eye to see minute objects not visible to the naked eye. **Resolution** of a microscope is the ability to distinguish between two objects as separate entities. At low resolution only one object may be detected. At high resolution two distinct objects are visible. At high resolution the image of such a specimen would show considerable detail.

The light microscope has limited resolution (0.2 μm) due to the wavelength of light so that organelles such as mitochondria, although visible, do not have clarity. Electron microscopes have exceptional resolution. The transmission electron microscope has a high resolution (0.2 to 0.3 nm). This enables even tiny organelles to be seen.

The light microscope

This type of microscope uses white light to illuminate a specimen. The light is focused onto the specimen by a **condensing lens**. The specimen is placed on a microscope slide which is clipped onto a platform, known as the **stage**. The image is viewed via an eyepiece or ocular lens. Overall magnification of the specimen depends on the individual magnification of the eyepiece lens and objective lens. For example, if a specimen is being observed with an eyepiece × 10 and an objective lens of × 40, then the image is 400 times the true size of the specimen.

> As the resolution of the light microscope is limited the best practical magnification is about 1500x. The TEM can magnify up to 300,000x

Many specimens need **staining** with chemicals so that tissues and, perhaps, organelles can be seen clearly, e.g. methylene blue is used to stain nuclei. Sometimes more than one stain is used, e.g. in differential staining, so that sub-cellular parts contrast with each other.

KEY POINT

> To decide which microscope is suitable use the table below.

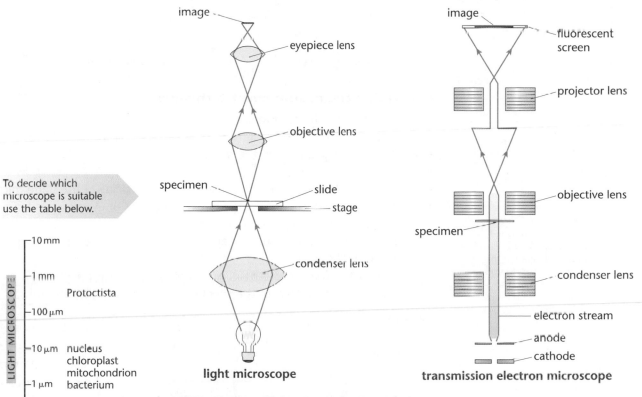

light microscope **transmission electron microscope**

ELECTRON MICROSCOPE | LIGHT MICROSCOPE

- 10 mm
- 1 mm — Protoctista
- 100 μm
- 10 μm — nucleus, chloroplast, mitochondrion
- 1 μm — bacterium
- 100 nm — vesicle, virus, ribosome
- 10 nm — cell membrane
- 1 nm
- 0.1 nm

The electron microscope

This uses an electron stream which is directed at the specimen. The **transmission electron microscope (TEM)** has extremely high magnification and resolution properties. Specimens are placed in a vacuum within the microscope, to ensure the electrons do not collide with air molecules and distort the image. **Stains** such as **osmium** and **uranium** salts are used to make organelles distinct. These salts are absorbed by organelles and membranes differentially, e.g. the nuclear membrane absorbs more of the salts than other parts. In this way the nuclear membrane becomes more dense. When the electron beam hits the specimen, electrons are unable to pass through this dense membrane. The membrane shows up as a dark shadow area on the image, because it is in an electron shadow. Cytoplasm allows more electrons to pass through. When these electrons hit the fluorescent screen visible light is emitted. The scanning electron microscope can achieve almost the same resolution as the TEM. It is used to see surface detail.

Artefacts

When microscopic specimens are prepared there are often several chemical and physical procedures. Often the material is dead so changes from the living specimen are expected. Microscopic material should be analysed with care because there may have been some artificial change in the material during preparation, e.g. next to some cells a student might see a series of small circles. They look like eggs but are merely air bubbles. These are **artefacts**; structures alien to the material which should not be interpreted as part of the specimen.

Microscopic measurement

It is sometimes necessary to measure microscopic structures. There are two instruments needed for this process, a **graticule** and **stage micrometer**.

graticule   stage micrometer

Calibration and measurement technique

- Put a graticule into the eyepiece of a microscope.
- Look through the eyepiece lens and the graticule line can be seen.
- Put a stage micrometer on the microscope stage.
- Look through the eyepiece lens.
- Line up the ruled line of the graticule with the ruled line of the stage micrometer.
- Calibrate the eyepiece by finding out the number of eyepiece units (e.u.) equal to one stage unit (s.u.); each is 0.01mm.
- If three eyepiece units equal 1 stage unit, then 1 eyepiece unit is equal to 0.01/3 (0.0033 mm).
- Take away the stage micrometer and replace it with a specimen.
- Measure the dimensions of the specimen in terms of eyepiece units.

> Important! Every time you change the objective lens, e.g. move from low power to high power, recalibration is necessary. In an examination this may be tested. Many candidates forget to recalibrate. Don't miss the mark!

Example

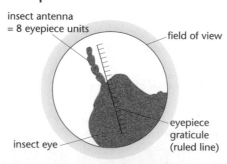

insect antenna = 8 eyepiece units

field of view

eyepiece graticule (ruled line)

insect eye

1 s.u. = 0.01 mm

3 e.u. = 1 s.u.

therefore 1 e.u. = 0.0033 mm

length of insect

antenna = 8 e.u.

= 8 × 0.0033 mm

= 0.0264 mm

Progress check

1 Describe how the width of an insect's leg could be measured using a microscope, graticule and stage micrometer.

2 A student measured the insect leg but considered that a higher magnification was needed to improve accuracy. What is the significance of changing magnification to the measuring technique?

2 Recalibration is needed for each new magnification.

1 Put a graticule into the eyepiece of a microscope; put a stage micrometer on the microscope stage; look through the eyepiece lens; line up the ruled line of the graticule with the ruled line of the stage micrometer; calibrate the eyepiece by finding out the number of eyepiece units equal to one stage unit; a stage unit is a known length, so an eyepiece unit length can be calculated; take away the stage micrometer and replace with specimen; measure width of insect leg in eyepiece units.

2.3 Specialisation of cells

After studying this section you should be able to:

- understand cell specialisation and how cells aggregate into tissues and organs
- recall a range of cell adaptations

LEARNING SUMMARY

The earlier parts of this chapter described the structure and function of generalised animal and plant cells. Their features are found in many unicellular organisms where all the life-giving processes are carried out in one cell. Additionally many multicellular organisms exist. A few show no specialisation and consist of repeated identical cells, e.g. Volvox, a colonial alga. Most multicellular organisms exhibit **specialisation**, where different cells are adapted for specific roles.

Cell adaptations

OCR   1.1.3

Some important features:

Red blood cell
- no nucleus
- high surface area
- contains haemoglobin which has an affinity for oxygen

red blood cell

White blood cell (neutrophil)
- can surround foreign cells and debris
- produces enzymes to digest the foreign material

neutrophil

Epithelial cell (e.g. squamous cell)
- very thin
- allows exchange of chemicals

epithelial cell

Plant palisade cell
- chloroplasts for photosynthesis
- chloroplasts can move to absorb more light
- contains chlorophyll which absorbs light
- sap vacuole stores important chemicals

palisade, mesophyll cell of leaf

Sperm cell
- flagellum for swimming
- numerous mitochondria for ATP production
- an acrosome containing enzymes to digest way into female gamete

sperm cell

Root hair cell
- projection of cell to increase surface area for water absorption and anchorage

root hair cell

Guard cell
- unevenly thickened cell walls so cells bend when turgid
- chloroplasts to provide energy for uptake of minerals and hence water.

guard cells

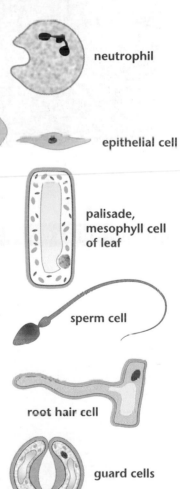

Tissues, organs and systems

OCR 1.1.3

A **tissue** is a collection of similar cells, derived from the same source, all working together for a specific function, e.g. palisade cells of the leaf which photosynthesise or the smooth muscle cells of the intestine which carry out peristalsis.

An **organ** is a collection of tissues which combine their properties for a specific function, e.g. the stomach includes: smooth muscle, epithelial lining cells, connective tissue, etc. Together they enable the stomach to digest food.

A range of tissues and organs combine to form a **system**, e.g. the respiratory system.

In multicellular organisms specific groups of cells are specialised for a particular role. This increased efficiency helps the organism to have better survival qualities in the environment.

The photomicrograph below shows some of the cells which are part of a bone.

Haversian canal containing blood vessels and nerves

Bone cells which secrete the minerals which harden the bone

Combinations of cells each contribute their specific adaptations to the overall function of an organ. Compact bone, spongy bone and articular cartilage all have distinct but vital qualities.

Prokaryotic cells

OCR 1.1.1

All the cells that have been described so far are called **eukaryotic** cells. All animals and plants are eukaryotic. Organisms such as bacteria are called **prokaryotic**. Their cells are much simpler. They do not have specialised organelles such as a nucleus or mitochondria. They contain a circular ring of DNA in the cytoplasm and their ribosomes are smaller in size.

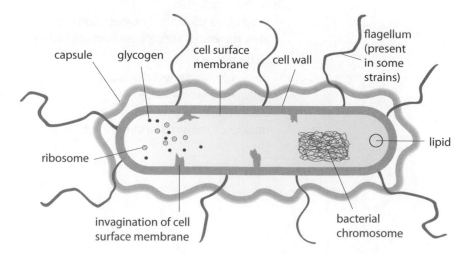

Sample question and model answer

The electron micrograph below shows a lymphocyte which secretes antibodies. Antibodies are proteins.

Analyse electron micrographs carefully. Learn the typical structures of all the organelles and then you will be prepared.

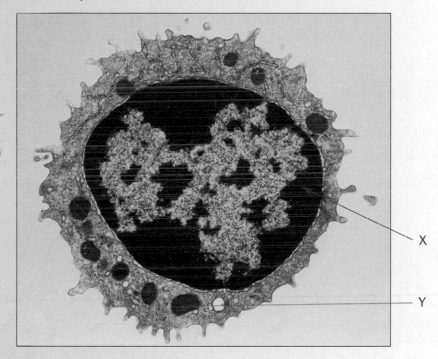

(a)

(i) Name the organelles X and Y. [2]

X = nucleus
Y = mitochondrion

(ii) The cell was stained with uranium salts in preparation for a transmission electron microscope. Explain how this stain caused the nucleus to show a dark shade compared to the light shade of the cytoplasm. [4]

The stain was taken up by the nucleus more than the cytoplasm; the electrons could not pass through the stained (dense) parts of the nucleus so the dark nucleus parts on the screen are in electron shade. Electrons pass through the cytoplasm and cause light emission (fluorescence) at the screen.

There are many different stains. You need to know the principle of how they work, but not specific names.

(b) The lymphocyte is specialised for the production and release of antibodies. Describe the role of the following organelles in this process.

(i) Nucleus [2]

Contains DNA; this contains the gene for the production of the antibody: this is where a mRNA molecule is formed.

All three parts to this question carry two marks each but there are three marking points for each.

(ii) Ribosomes [2]

The ribosome is the site of protein synthesis / production of the antibody. The mRNA attaches to the ribosome. tRNA brings the correct amino acids into line for forming the protein.

(iii) Golgi body [2]

Vesicles containing the antibodies join with the Golgi body. The antibodies may be modified in the Golgi. The antibody is then packaged for secretion.

Practice examination questions

1 The diagram shows the structure of a cell surface membrane.

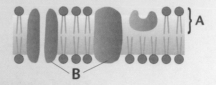

(a) Name molecules A and B. [2]

(b) Describe one role of molecule B. [1]

(c) Explain why the structure of molecule A means that the membrane forms a bilayer. [2]

2 The diagram below shows an electron micrograph of a cell.

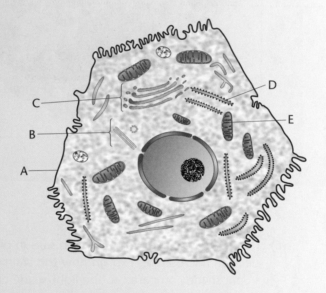

(a) Name the parts labelled **A**, **B**, **C**, **D** and **E**. [5]

(b) The magnification of this diagram is 10 000. Work out the actual diameter of the nucleus. Give your answer in micrometers. [3]

(c) The cell is a liver cell. It contains many mitochondria.

Explain why there are so many mitochondria in each liver cell. [2]

Practice examination questions (continued)

3   The diagram below shows the structure of a transmission electron microscope (TEM).

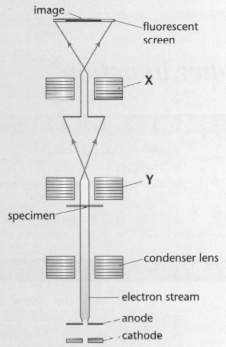

transmission electron microscope

(a) Name lens X and lens Y.   [2]

(b) Why is it necessary for the specimen to be put in a vacuum?   [1]

(c) Occasionally an image seen when using the electron microscope shows an item not present in the living organism.

(i)   What name is given to this type of item?   [1]

(ii) How should the presence of the item be interpreted?   [1]

Chapter 3
Enzymes

The following topics are covered in this chapter:

- *Enzymes in action*
- *Inhibition of enzymes*

3.1 Enzymes in action

After studying this section you should be able to:

- *understand the role of the active site and the enzyme–substrate complex in enzyme action*
- *understand how enzymes catalyse biochemical reactions by lowering activation energy*
- *understand the factors which affect the rate of enzyme catalysed reactions*

LEARNING SUMMARY

How enzymes work

OCR 2.1.3

Living cells carry out many biochemical reactions. These reactions take place **rapidly** due to the presence of enzymes. All enzymes consist of **globular proteins** which have the ability to 'drive' biochemical reactions. Some enzymes require additional non-protein groups to enable them to work efficiently, e.g. the enzyme dehydrogenase needs a coenzyme NAD to function. Most enzymes are contained within cells but some may be released and act extracellularly.

> The tertiary folding of polypeptides are responsible for the special shape of the active site.

> **The ability of an enzyme to function depends on the specific shape of the protein molecule. The intricate shape created by polypeptide folding (see page 22) is a key factor in both theories of enzyme action.**
>
> **KEY POINT**

Lock and key theory

> In an examination the lock and key theory is the most important model to consider. Remember that both catabolic and anabolic reactions may be given.

- Some part of the enzyme has a cavity with a precise shape (**active site**).
- A substrate can fit into the active site.
- The active site (lock) is exactly the correct shape to fit the substrate (key).
- The substrate binds to the enzyme forming an **enzyme–substrate complex**.
- The reaction takes place rapidly.
- Certain enzymes break a substrate down into two or more products (**catabolic** reaction).
- Other enzymes bind two or more substrates together to assemble one product (**anabolic** reaction).

> metabolic reaction
> = anabolic + catabolic
>   reaction   reaction
> Remember that metabolism is a summary of **build up** and **break down reactions**.

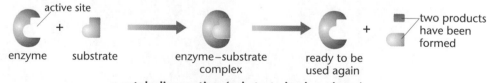

a catabolic reaction (substrate broken down)

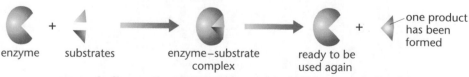

an anabolic reaction (substrates used to build a new molecule)

Induced fit theory

- The active site is a cavity of a particular shape.
- Initially the active site is not the correct shape in which to fit the substrate.
- As the substrate approaches the active site, the site changes and this results in it being a perfect fit.
- After the reaction has taken place, and the products have gone, the active site returns to its normal shape.

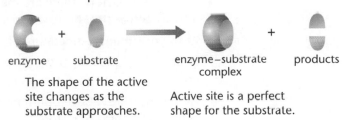

enzyme   substrate   enzyme–substrate   products
complex

The shape of the active site changes as the substrate approaches.

Active site is a perfect shape for the substrate.

Lowering of activation energy

Every reaction requires the input of energy. Enzymes reduce the level of activation energy needed as shown by the graph.

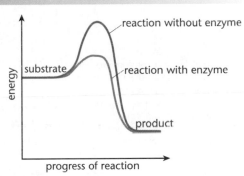

The higher the activation energy the slower the reaction. An enzyme reduces the amount of energy required for a biochemical reaction. When an enzyme binds with a substrate the available energy has a greater effect and the rate of catalysis increases. The conditions which exist during a reaction are very important when considering the rate of progress. Each of the following has an effect on the rate:

- concentration of substrate molecules
- concentration of enzyme molecules
- temperature
- pH.

You may be questioned on the factors which affect the rate of reaction. Less able candidates tend to remember just one or two factors. Learn all four factors here and achieve a higher grade!

Look out for questions which show the rate of reaction graphically. The examiners often test your understanding of limiting factors (see practice question on page 46).

What is the effect of enzyme and substrate concentration?

When considering the rate of an enzyme catalysed reaction the proportion of enzyme to substrate molecules should be considered. Every substrate molecule fits into an active site, then the reaction takes place. If there are more substrate molecules than enzyme molecules then the number of active sites available is a **limiting factor**. The **optimum rate** of reaction is achieved when all the active sites are in use. At this stage if more substrate is added, there is no increase in rate of product formation. When there are fewer substrate molecules than enzyme molecules the reaction will take place very quickly, as long as the conditions are appropriate.

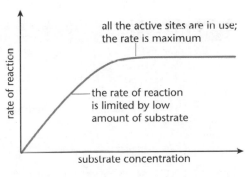

all the active sites are in use; the rate is maximum

the rate of reaction is limited by low amount of substrate

Remember that particles in liquids (and gases) are in constant random motion, even though we cannot see them.

How does temperature affect the rate of an enzyme catalysed reaction?

- **Heat energy** reaching the enzyme and substrate molecules causes them to **increase random movement**.

- The greater the heat energy the more the molecules move and so the more often they **collide**.

- The more **collisions** there are the greater the chance that substrates will fit into an **active site**, up to a specific temperature.

- At the **optimum** temperature of an enzyme, the reaction rate is maximum.

- Heat energy also affects the shape of the active site, the active site being the correct shape at the optimum temperature.

- At temperatures above optimum, the rate of reaction decreases because the active site begins to distort.

- Very high temperature causes the enzyme to become **denatured**, i.e. bonding becomes irreversibly changed and the active site is **permanently damaged**.

- At very high temperatures, the number of collisions is correspondingly high, but without active sites no products can be formed.

- At lower temperatures than the optimum, the rate of the reaction decreases because of reduced enzyme–substrate collisions.

It is interesting to consider that some micro organisms can spoil ice-cream in a freezer whereas a different micro organism, with different enzymes, can decompose grass in a 'steaming' compost heap.

Most enzymes have an optimum temperature of between 30°C and 40°C, but there are many exceptions. An example of this is shown by some bacteria that live at high temperatures in hot volcanic springs.

How does pH affect the rate of an enzyme catalysed reaction?

The **pH** of the medium can have a direct effect on the bonding responsible for the **secondary and tertiary structure** of enzymes. If the active site is changed then enzyme action will be affected. Each enzyme has an optimum pH.

Remember that other factors affect an enzyme catalysed reaction:

- substrate concentration
- enzyme concentration
- temperature.

Each can be a limiting factor.

- Many enzymes work best at **neutral** or **slightly alkaline** conditions, e.g. salivary amylase.

- Pepsin works best in **acid** conditions around pH 3.0, as expected considering that the stomach contains hydrochloric acid.

For the two enzyme examples above, the active sites are ideally shaped at the pHs mentioned. An inappropriate pH, often acidic, can change the active site drastically, so that the substrate cannot bind. The reaction will not take place. On most occasions the change of shape is not permanent and can be returned to optimum by the addition of an alkali.

Progress check

How does temperature affect the rate of the reaction by which protein is changed to polypeptides by the enzyme pepsin, in the human stomach?

$$\text{protein} \xrightarrow{\text{pepsin}} \text{polypeptides}$$

- Heat energy causes the enzyme and substrate molecules to increase random movement, increasing the chance of collision.
- At 37°C (optimum temperature) there is a greater chance that the protein will fit into an active site, so the production of polypeptides is at maximum rate.
- At 37°C (optimum temperature) the shape of the active site is best suited to fit the protein.
- At temperatures higher than 37°C the rate of reaction decreases because the active site begins to distort.
- Very high temperature causes the pepsin to become denatured, i.e. bonding has been irreversibly changed and the active site is permanently damaged.
- At very high temperatures the number of collisions is correspondingly high, but without active sites no polypeptides can be formed.
- At lower temperatures than optimum the rate of reaction decreases because of reduced enzyme–substrate collisions.

3.2 Inhibition of enzymes

After studying this section you should be able to:

- *understand the action and effects of competitive and non-competitive inhibitors*

LEARNING SUMMARY

What are inhibitors?

OCR 2.1.3

Certain chemicals can slow down or stop enzyme catalysed reactions. These chemicals are called **inhibitors**.

Sometimes these chemicals are substances that are naturally occurring inside cells. They may be used to regulate the rate of enzyme controlled reactions. Other inhibitors may be poisons or medicines.

Competitive inhibitors

- These are molecules of **similar shape** to the normal substrate and are able to bind to the active site.
- They do not react within the active site, but leave after a time without any product forming.
- The enzymic reaction is **reduced** because while the inhibitor is in the active site, **no substrate can enter**.
- Substrate molecules **compete** for the active site so the rate of reaction decreases.
- The higher the proportion of competitive inhibitor the slower the rate of reaction.

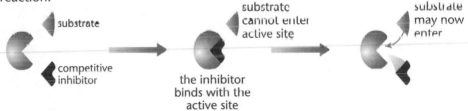

substrate

competitive inhibitor

the inhibitor binds with the active site

substrate cannot enter active site

substrate may now enter

Some enzymes have two sites, the active site and one other. An **allosteric** molecule fits into the alternative site. Here it changes the shape of the active site. This can stimulate the reaction if the active site becomes a better shape (**allosteric activation**). It can also inhibit if the active site becomes an inappropriate shape (**allosteric inhibition**).

Non-competitive inhibitors

- These are molecules which bind to some part of an enzyme other than the active site.
- They have a different shape to the normal substrate.
- They change the shape of the active site which no longer allows binding of the substrate.
- Some substrate molecules may reach the active site before the non-competitive inhibitor.
- The rate of reaction is reduced.
- Finally they leave their binding sites, but substrate molecules do not compete for these, so they have a greater inhibitory effect.

non-competitive inhibitor

substrate

binding site

active site has changed

substrate has opportunity to enter

The graph below shows the relative effects of competitive and non-competitive inhibitors, compared to a normal enzyme catalysed reaction.

Many poisons act as enzyme inhibitors, such as cyanide. Many drugs are now made that also target enzymes.

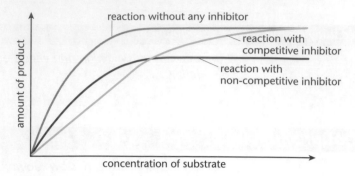

Sometimes an inhibitor will not leave the enzyme once it has bound with it. It stays permanently attached. This is called an **irreversible inhibitor**. Inhibitors that do leave are called **reversible**.

Progress check

1 Explain why a non-competitive inhibitor does not need to have the same shape as the substrate. Give a reason for your answer.

2 How does a non-competitive inhibitor reduce the rate of an enzyme catalysed reaction?

1 No. Non-competitive inhibitors have a different shape to the normal substrate. They bind to some part of an enzyme other than the active site.

2 They change the shape of the enzyme's active site which is less suitable for the binding of the substrate, so the rate of reaction decreases.

Sample question and model answer

The graph below shows the rate of an enzyme catalysed reaction, with and without a **non-competitive inhibitor**, at different substrate concentrations.

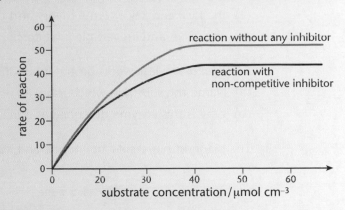

(a) Give **one** piece of evidence from the graph which shows that the inhibitor was non-competitive rather than competitive. [1]

When a non-competitive inhibitor is present the maximum rate of the inhibitor-free reaction is not reached.

> A competitive inhibitor would still allow the maximum rate at high substrate concentration.

(b) Explain why the rates of the inhibited and non-inhibited reactions were very similar up to a substrate concentration of 20 μmol cm^{-3}. [2]

Before a concentration of 20 μmol cm^{-3}
- *in both reactions substrate molecules are similarly successful in reaching active sites*
- *few inhibitor molecules have become bound to the alternative sites on the enzyme*
- *few active sites have been changed so are still available for the substrate.*

> Remember that the non-competitive inhibitor binds to a different part of an enzyme and NOT the active site. It still causes a change in the active site which cannot then bind with the substrate.

(c) Give **two** factors that are needed to be kept constant when investigating both the inhibited and non-inhibited reactions. [2]

- *temperature*
- *pH*

Practice examination questions

1 Describe each of the following pairs to show that you understand the main differences between them:

(a) the lock and key enzyme theory **and** the induced fit enzyme theory

(b) reversible **and** irreversible inhibitors. [4]

2 Bacterial α-amylase works best at around 80°C.

(a) Name the substrate which it breaks down. [1]

(b) Why is this enzyme described as a thermostable enzyme? [1]

3 The diagram represents an enzyme and its substrate.

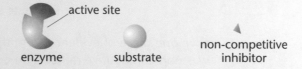

active site

enzyme   substrate   non-competitive inhibitor

(a) Referring to information in the diagram explain the activity of this enzyme in terms of the **induced fit theory**. [2]

(b) Explain how the non-competitive inhibitor has an effect on the structure and function of the enzyme. [3]

4 The diagram below shows a long polypeptide.

polypeptide   O–H

(a) The carboxylic acid is found at one end of the polypeptide. Which group is found at the other end? [1]

(b) Pepsin breaks down polypeptides by breaking bonds in the centre of the molecule. This speeds up the subsequent digestion of the polypeptide. Explain how. [2]

Chapter 4

Exchange

The following topics are covered in this chapter:

- The cell surface membrane
- The movement of molecules in and out of cells
- Gaseous exchange in humans

4.1 The cell surface membrane

After studying this section you should be able to:

- understand the importance of surface area to volume ratio
- recall the fluid mosaic model of the cell surface (plasma) membrane

How important is the surface area of exchange surfaces?

OCR   1.2.1

Unicellular organisms like amoeba have a very high **surface area to volume ratio**. All chemicals that are needed can pass into the cells directly and all waste can pass out efficiently. Organisms which have a high surface area to volume ratio have no need for special structures like lungs or gills.

Nutrients and oxygen passing into an organism are rapidly used up. This limits the ultimate size to which a unicellular organism can grow. If vital chemicals did not reach all parts of a cell then death would be a consequence.

A unicellular organism may satisfy all its needs by direct diffusion. However, in larger organisms cells join to adjacent ones and surfaces exposed for exchange of substances are reduced. The larger an organism the lower its surface area to volume ratio. For this reason many multicellular organisms have specially adapted exchange structures.

Fluid mosaic model of the cell surface (plasma) membrane

OCR   1.1.2

Ultimately the exchange of substances takes place across the cell surface membrane. This must be selective, allowing some substances in and excluding others. The cell membrane consists of a bilayer of phospholipid molecules (see page 20-21). Each phospholipid is arranged so that the hydrophilic (attracts water) head is facing towards either the cytoplasm or the outside of the cell. The hydrophobic (repels water) tails meet in the middle of the membrane. Across this expanse of phospholipids are a number of protein molecules. Some of the proteins (intrinsic) span the complete width of the membrane, some proteins (extrinsic) are partially embedded in the membrane.

Remember that plant cells have a cellulose cell wall. This gives physical support to the cell but is permeable to many molecules. Water and ions can readily pass through.

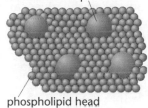

upper surface of cell membrane protein

phospholipid head

The fluid mosaic model of the cell membrane

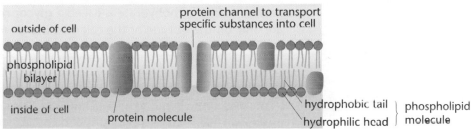

outside of cell

protein channel to transport specific substances into cell

phospholipid bilayer

inside of cell

protein molecule

hydrophobic tail
hydrophilic head

phospholipid molecule

Functions of cell membrane molecules

The term 'fluid mosaic' was given to the cell membrane because of the dynamic nature of the component molecules of the membrane. Many of the proteins seem to 'float' through an apparent 'sea' of phospholipids. Few molecules are static. The fluidity of the membrane is controlled by the quantity of cholesterol molecules. These are found inbetween the tails of the phospholipids.

Phospholipid

Small lipid-soluble molecules pass through the membrane easily because they dissolve as they pass through the phospholipid's bilayer. Small uncharged molecules also pass through the bilayer.

small lipid-soluble molecules pass through

Channel proteins (ion gates)

Larger molecules and charged molecules can pass through the membrane due to channel proteins. Some are adjacent to a receptor protein, e.g. at a synapse a transmitter substance binds to a receptor protein. This opens the channel protein or ion gate and sodium ions flow in.

Not all channel proteins need a receptor protein.

> When the molecule binds to a receptor molecule it is similar to a substrate binding with an enzyme's active site. On this occasion the receptor site is the correct shape.

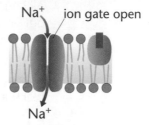

transmitter substance

receptor protein

Na⁺ ion gate open

Na⁺

Carrier protein molecule

Some molecules which approach a cell may bind with a carrier protein. This has a site which the incoming molecule can bind to. This causes a change of shape in the carrier protein which deposits the molecule into the cell cytoplasm.

once in position the molecule changes the shape of the carrier protein

the site gives up the molecule on the inside of the cell

carrier protein

Recognition proteins

These are extrinsic proteins, some having carbohydrate components, which help in cell recognition (cell signalling) and cell interaction, e.g. foreign protein on a bacterium would be recognised by white blood cells and the cell would be attacked. The combination of a protein with a carbohydrate is called a **glycoprotein**. The carbohydrate chains are only on the outside of the cell membrane and are called the **glycocalyx**.

> White blood cells continually check the proteins on cell membranes. Those recognised as 'self' are not attacked, whereas those which are not 'self' are attacked.

> The proteins may act as receptor sites for hormones or drugs

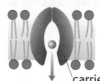

carbohydrate

recognition protein

> **KEY POINT**
>
> The cell surface membrane is the key structure which forms a barrier between the cell and its environment. Nutrients, water and ions must enter and waste molecules must leave. Equally important is the exclusion of dangerous chemicals and inclusion of vital cell contents. It is no surprise that the cell makes further compartments within the cell using membranes of similar structure to the cell surface membrane. High temperatures can destroy the structure of the cell membrane and then it loses its ability to contain the cell contents.

4.2 The movement of molecules in and out of cells

After studying this section you should be able to:

- *understand the range of methods by which molecules cross cell membranes*
- *understand the processes of diffusion, facilitated diffusion, osmosis and active transport*

LEARNING SUMMARY

How do substances cross the cell surface membrane?

OCR ▶ 1.1.2

Cells need to obtain substances vital in sustaining life. Some cells secrete useful substances but all cells excrete waste substances. There are several mechanisms by which molecules move across the cell surface membrane.

Diffusion

Note that diffusion is the movement of molecules down a concentration gradient.

Molecules in liquids and gases are in constant random motion. When different concentrations are in contact, the molecules move so that they are in equal concentration throughout. An example of this is when sugar is put into a cup of tea. If left, sugar molecules will distribute themselves evenly, even without stirring. Diffusion is the movement of molecules from where they are in high concentration to where they are in low concentration. Once evenly distributed the *net* movement of molecules stops.

Factors which affect the rate of diffusion

- Surface area.
  the greater the surface area the greater the rate of diffusion
- The difference in concentration on either side of the membrane.
  the greater the difference the greater the rate
- The size of molecules.
  smaller molecules may pass through the membrane faster than larger ones
- The presence of pores in the membrane.
  pores can speed up diffusion
- The width of the membrane.
  the thinner the membrane the faster the rate.

Sometimes the membrane is stated as being selectively permeable, partially permeable or semi-permeable. They all mean the same.

Facilitated diffusion

This is a special form of diffusion in which protein carrier molecules are involved. It is much faster than regular diffusion because of the carrier molecules. Each carrier will only bind with a specific molecule. Binding changes the shape of the carrier which then deposits the molecule into the cytoplasm. No energy is used in the process.

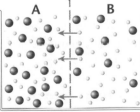

selectively permeable
membrane

A   B

○ water molecule
● solute molecule

water molecules move
from B to A

Osmosis

This is the movement of water molecules across a selectively permeable membrane:

- from a lower concentrated solution to a higher concentrated solution
- from where water molecules are at a higher concentration to where they are at a lower concentration
- from a hypotonic solution to a hypertonic solution
- from a hyperosmotic solution to a hypo-osmotic solution
- from an area of higher water potential to lower water potential.

The diagram on the left shows a model of osmosis.

Remember that osmosis is about the movement of **water molecules**. No other substance moves!

Animal cells act differently to plant cells because they do not have a cell wall. They cannot become turgid, they just burst or lyse if they take up too much water.

What is the relationship between water potential of the cell and the concentration of an external solution?

The term 'water potential' is used as a measure of water movement from one place to another in a plant. It is measured in terms of pressure and the units are either kPa (kilopascals) or MPa (megapascals). Water potential is indicated by the symbol ψ (*pronounced psi*). The following equation allows us to work out the water 'status' of a plant cell.

ψ (cell) water potential (of cell)	$=$	ψs solute potential (of ions inside cell)	$+$	ψp pressure potential (of cell wall)	**KEY POINT**

Remember that water moves from an area of higher water potential to an area of lower water potential. When a cell at −4 MPa is next to a cell at the less negative value the water moves to the more negative value, i.e. −4 MPa > −6 MPa

Note that pressure potential only has a value **above** zero when the cell membrane **begins** to contact the cell wall. The greater the pressure potential the more the cell wall resists water entry. At turgidity ψs = ψp when net water movement is zero.

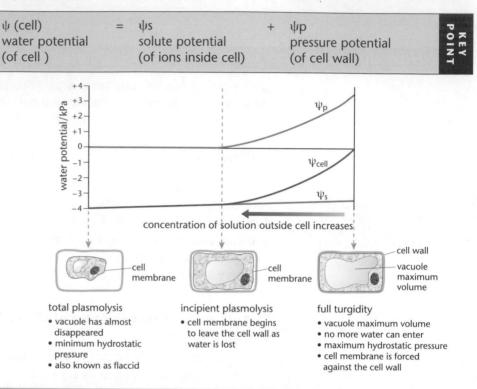

concentration of solution outside cell increases

total plasmolysis
- vacuole has almost disappeared
- minimum hydrostatic pressure
- also known as flaccid

incipient plasmolysis
- cell membrane begins to leave the cell wall as water is lost

full turgidity
- vacuole maximum volume
- no more water can enter
- maximum hydrostatic pressure
- cell membrane is forced against the cell wall

Active transport

Note that active transport is the movement of molecules up a concentration gradient.

In **active transport** molecules move from where they are in lower concentration to where they are in higher concentration. A protein carrier molecule is used. This is **against the concentration gradient** and always **needs energy**. A plant may contain a higher concentration of Mg^{2+} ions than the soil. It obtains a supply by active transport through the cell surface membranes of root hairs. Only Mg^{2+} ions can bind with the specific protein carrier molecules responsible for their entry into the plant. This is also known as active ion uptake, but is a form of active transport. Any process that reduces respiration in cells will reduce active transport, e.g. adding a poison such as cyanide or reducing oxygen availability. This is because energy is needed for the process and this energy is released by respiration.

In the small intestine, glucose is absorbed into the bloodstream indirectly by active transport. Sodium ions are pumped out of the epithelial cells allowing glucose and sodium ions to diffuse into the cell by a co-transporter system.

Endocytosis, exocytosis, pinocytosis and phagocytosis

Some substances, often due to their large size, enter cells by **endocytosis** as follows:

- the substance contacts the cell surface membrane which indents
- the substance is surrounded by the membrane, forming a vacuole or vesicle
- each vacuole contains the substance and an outer membrane which has detached from the cell surface membrane.

When fluids enter the cell in this way this is known as **pinocytosis**. When the substances are large solid particles, this is called **phagocytosis**. Some substances

endocytosis

vacuole

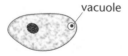

leave the cell in a reverse of endocytosis. Here the membrane of the vacuole or vesicle merges with the cell surface membrane depositing its contents into the outside environment of the cell. This is known as **exocytosis**.

Progress check

1 A plant contains a greater concentration of Fe^{2+} ions than the soil in which it is growing. Name and describe the process by which the plant absorbs the ions against the concentration gradient.

2 Explain the following:
(a) Endocytosis of an antigen by a phagocyte
(b) Exocytosis of amylase molecules from a cell.

amylase contents deposited outside of the cell.
(b) **Exocytosis:** a vesicle in the cell contains amylase molecules; the vesicle merges with the cell membrane;
vacuole; the vacuole contains the antigen and an outer membrane which has detached from the cell surface membrane.
2 (a) **Endocytosis:** antigen contacts the cell membrane of the phagocyte; cell membrane surrounds the antigen, forming a
which allow entry into the plant.
root hairs; protein carrier molecules in membranes used; energy needed; Fe^{2+} ions can bind with the protein carrier molecules
1 **Active transport:** molecules move from a lower concentration to a higher concentration; through the cell surface membranes of

4.3 Gaseous exchange in humans

After studying this section you should be able to:

- understand why organisms need to be adapted for gaseous exchange
- explain how the human breathing system brings about ventilation and gaseous exchange

LEARNING SUMMARY

How are organisms adapted for efficient gaseous exchange?

OCR 1.2.1

The range of respiratory surfaces in this chapter each have common properties, such as high surface area to volume ratio, one cell thick lining tissue, many capillaries.

The exchange of substances across cell surface membranes has been described. Larger organisms have a major problem in exchange because of their low surface area to volume ratio. Some organisms, like flatworms, have a large surface area due to the shape of their body. Other organisms satisfy their needs by having tissues and organs which have special adaptations for efficient exchange. In simple terms, these structures achieve a very high surface area, e.g. a leaf, and link to the transport system to allow import and export from the organ.

Gaseous exchange in humans

OCR 1.2.1

The diagram below shows the **human gas exchange system**. The alveoli are the site of gaseous exchange and they are connected to the outside air via a system of branching tubes.

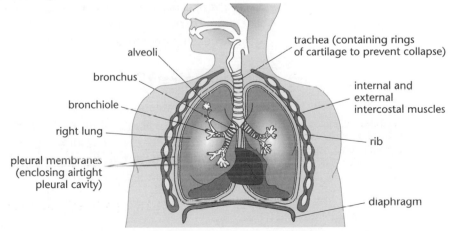

trachea (containing rings of cartilage to prevent collapse)

alveoli

bronchus

bronchiole

internal and external intercostal muscles

right lung

rib

pleural membranes (enclosing airtight pleural cavity)

diaphragm

The airways leading to the lungs have a number of different tissues that carry out different functions.

The trachea and bronchi contain rings of **cartilage** that help to keep the tubes open when the pressure inside drops.

Smooth muscle in the walls of the tubes can contract and relax to change the diameter of the airway.

Elastic tissue in the wall of the smaller airways helps to force the air out during exhalation due to elastic recoil.

Goblet cells will produce mucus which will trap particles. The **cilia** then waft this mucus up to the mouth to be swallowed.

Ventilation

Drawing air in and out of the lungs involves changes in pressure and volume in the chest. These changes work because the **pleural** membranes form an airtight **pleural cavity**.

Breathing in (inhaling):
1. The external intercostal muscles contract, moving the ribs upwards and outwards.
2. The diaphragm contracts and flattens.
3. Both of these actions will increase the volume in the pleural cavity and so decrease the pressure.
4. Air is therefore drawn into the lungs.

Breathing out (exhaling):
1. The internal intercostal muscles relax and the ribs move down and inwards.
2. The diaphragm relaxes and domes upwards.
3. The volume in the pleural cavity is decreased so the pressure is increased.
4. Air is forced out of the lungs.

Lungs of a mammal

The ventilation mechanism allows inhalation of air, which then diffuses into alveoli to exchange the respiratory gases. Completion of ventilation takes place when gases are expelled into the atmosphere. The diagram on the left shows the structure of the alveoli.

Adaptations of lungs for gaseous exchange

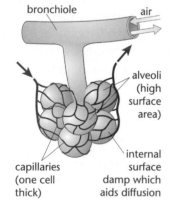

bronchiole

air

alveoli (high surface area)

internal surface damp which aids diffusion

capillaries (one cell thick)

* Air flows through a trachea (windpipe) supported by cartilage.
* It reaches the alveoli via tubes known as bronchi and bronchioles.
* Lungs have many alveoli (air sacs) which have a high surface area.
* Each alveolus is very thin (diffusion is faster over a short distance).
* Each alveolus has an inner film of moisture containing a chemical called surfactant. This reduces the surface tension and makes it easier to inflate the lungs.
* Each alveolus has many capillaries, each one cell thick, to aid diffusion.
* There are many blood vessels in the lungs to give a high surface area for gaseous exchange and transport of respiratory substances.

Measuring ventilation

The **maximum** volume of air that can be breathed out in one breath is called the vital capacity.

The process of ventilation can be investigated using a device called a **spirometer**. It can measure the volume of air exchanged in a single breath. This is called the **tidal volume**. It can also measure the number of breaths per minute – the **breathing rate**. If these two figures are multiplied together, the result gives the volume of gas exchanged in one minute, the **pulmonary ventilation**.

pulmonary ventilation = tidal volume × breathing rate

Sample question and model answer

The diagram below shows two adjacent plant cells A and B.
The water potential of a cell is ψ(cell).

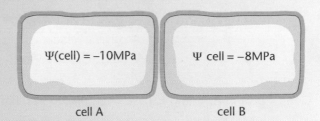

cell A   cell B

(a) (i)   Draw an arrow on the diagram to show the direction of water flow.
          Explain how you worked out the direction.   [2]

          The arrow should be drawn from cell B to cell A.
          Direction from −8 MPa to −10 MPa from a larger to a smaller ψ(cell).

    (ii) Name the condition of the cell when ψ(cell) = 0   [1]

          full turgor or fully turgid

(b) Give **one** difference between the following terms: [2]

    facilitated diffusion

    molecules move down a gradient

    active transport

    energy is needed for the process

Be careful with this type of question. You may believe that 'up a gradient' could be given for active transport. It is correct, but it's too close to the 'down a gradient' idea for facilitated diffusion.

Go for a completely different idea, as shown.

(c) What effect would the following temperatures have on the active transport
    of Mg^{2+} ions across a cell surface membrane of a plant cell? Assume the
    plant is a British native.   [4]

    (i)   30°C

          It is likely that active transport would be efficient because the
          temperature would be ideal for the Mg^{2+} to bind with a carrier
          molecule.

    (ii) 80°C

          process likely not to work;
          protein carrier denatured;
          Mg^{2+} would not be able to bind.

Practice examination questions

1

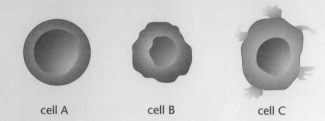

cell A   cell B   cell C

Cells A, B and C have been placed in different concentrations of salt solution.

(a) Explain each of the following in terms of water potential.

   (i)   Cell A did not change size at all.

   (ii) Cell B decreased in volume.

   (iii) Cell C became swollen and burst.   [3]

(b) Which process is responsible for the changes to cells B and C?   [1]

2 (a) Give one similarity between active transport and facilitated diffusion. [1]

   (b) Give one difference between active transport and facilitated diffusion. [1]

3 Describe and explain how the alveoli are adapted to efficient gaseous
exchange.   [4]

Chapter 5

Transport

The following topics are covered in this chapter:

- Mass transport systems
- Heart: structure and function
- Blood vessels
- The transport of substances in the blood
- The transport of substances in a plant

5.1 Mass transport systems

After studying this section you should be able to:

- explain why most multicellular organisms need a mass transport system
- understand the importance of a high surface area to volume ratio
- describe the differences between open and closed circulatory systems
- understand the implications of using single or double circulatory systems

LEARNING SUMMARY

Why do most multicellular organisms need a mass transport system?

OCR 1.2.2-3

The bigger an organism is, the lower its surface area to volume ratio. Substances needed by a large organism could not be supplied through its exposed external surface. Oxygen passing through an external surface would be rapidly used up before reaching the many layers of underlying cells. Similarly waste substances would not be excreted quickly enough. This problem has been solved, through evolution, by specially adapted tissues and organs.

> Leaves, roots, gills and lungs all have high surface area to volume properties so that supplies of substances vital to **all** the living cells are made available by these structures. Movement of substances to and from these structures is carried out by efficient **mass transport systems.**
>
> **KEY POINT**

This is a little like transport on trains where people travel together on the same train, in the same direction, at the same speed, but may get off at different places.

In a mass transport system, all the substances move in the same direction at the same speed. Across the range of multicellular organisms found in the living world are a number of mass transport systems, e.g. the mammalian circulatory system and the vascular system of a plant.

Mass transport systems are just as important for the rapid removal of waste as they are for supplies. Supplies include an immense number of substances, e.g. glucose, oxygen and ions. Even communication from one cell to another can take place via a mass transport system, e.g. hormones in a blood stream.

The greater the metabolic rate of an organism, the greater the demands on its mass transport system.

Different types of circulatory systems

OCR 1.2.2

Most organisms that are beyond a certain size have a circulatory system. These systems may be open or closed.

> In a **closed circulation** the blood is contained in blood vessels as it circulates. In an **open circulation** the blood is contained in the body cavity.
>
> **KEY POINT**

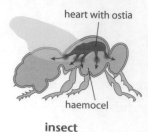
heart with ostia

haemocel

insect

Insects have an open circulation. The blood is in the body cavity called the haemocoel. It does not circulate in blood vessels but a dorsal tube-shaped heart maintains movement of the blood in the chamber.

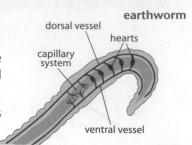
earthworm
dorsal vessel
hearts
capillary system
ventral vessel

Earthworms have a closed circulation. Five of the blood vessels act like hearts, pumping the blood through the main two blood vessels.

In vertebrates, the pumping of the blood is performed by a specialised heart.

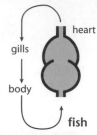
heart
gills
body
fish

> **KEY POINT**
> Vertebrate circulatory systems can be either single or double and this results in structural differences between their hearts.

Fish have a **single circulatory system**. This means that the blood flows through the heart once on each circuit of the body.

Mammals have a **double circulatory** system. This means that as blood enters the heart it is pumped to the lungs, exchanges carbon dioxide for oxygen, and returns to the heart where further pumping propels it through the rest of the body. The blood moves through the heart twice during each cardiac cycle.

> In a single circulation, the blood has to pass through two capillary beds, one after the other. This makes blood flow more slowly compared with a double circulation.

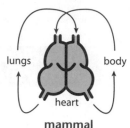
lungs
body
heart
mammal

5.2 Heart: structure and function

After studying this section you should be able to:

> **LEARNING SUMMARY**

- recall the structure, cardiac cycle and electrical stimulation of a mammalian heart

The mammalian heart

OCR 1.2.2

The heart consists of a range of tissues. The most important one is cardiac muscle. The cells have the ability to contract and relax through the complete life of the person, without ever becoming fatigued. Each cardiac muscle cell is **myogenic**. This means it has its own inherent rhythm. Below are diagrams of the heart and its position in the circulatory system.

> **Note** that tricuspid and bicuspid valves are known as atrioventricular valves.

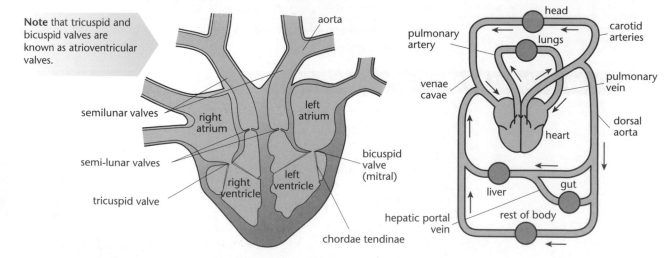

aorta
left atrium
right atrium
semilunar valves
semi-lunar valves
tricuspid valve
right ventricle
left ventricle
bicuspid valve (mitral)
chordae tendinae

head
carotid arteries
pulmonary artery
lungs
venae cavae
pulmonary vein
dorsal aorta
heart
liver
gut
hepatic portal vein
rest of body

Structure

The heart consists of four chambers, **right** and **left atria** above **right** and **left ventricles**. The functions of each part are as follows.

If blood moved in the wrong direction, then transport of important substances would be impeded.

- The **right atrium** links to the **right ventricle** by the **tricuspid valve**. This valve prevents backflow of the blood into the atrium above, when the ventricle contracts.

- The **left atrium** links to the **left ventricle** by the **bicuspid valve (mitral valve)**. This also prevents backflow of the blood into the atrium above.

- The **chordae tendonae** attach each ventricle to its **atrioventricular valve**. Contractions of the ventricles have a tendency to force these valves up into the atria. Backflow of blood would be dangerous, so the chordae tendonae hold each valve firmly to prevent this from occurring.

Check out these diagrams of a valve.

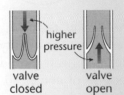

valve closed   valve open

higher pressure

You can work out if a valve is open or closed in terms of pressure. Higher pressure above than below a semi-lunar valve closes it. Higher pressure below the semi-lunar valve than above, opens it.

- Semi-lunar (pocket) valves are found in the blood vessels leaving the heart (pulmonary artery and aorta) They only allow exit of blood from the heart through these vessels following ventricular contractions. Elastic recoil of these arteries and relaxation of the ventricles closes each semi-lunar valve.

- Ventricles have thicker muscular walls than atria. When each atrium contracts it only needs to propel the blood a short distance into each ventricle.

- The left ventricle has even thicker muscular walls than the right ventricle. The left ventricle needs a more powerful contraction to propel blood to the systemic circulation (all of the body apart from the lungs). The right ventricle propels blood to the nearby lungs. The contraction does not need to be so powerful.

Cardiac cycle

Blood must continuously be moved around the body, collecting and supplying vital substances to cells as well as removing waste from them. The heart acts as a pump using a combination of **systole** (contractions) and **diastole** (relaxation) of the chambers. The cycle takes place in the following sequence.

Examination questions often test your knowledge of the opening and closing of valves. Always analyse the different pressures given in the question. A greater pressure behind a valve opens it. A greater pressure in front closes it.

Stage 1

Ventricular diastole, atrial systole
Both ventricles relax simultaneously. This results in lower pressure in each ventricle compared to each atrium above. The atrioventricular valves open partially. This is followed by the atria contracting which forces blood through the atrioventricular valves. It also closes the valves in the vena cava and pulmonary vein. This prevents backflow of blood.

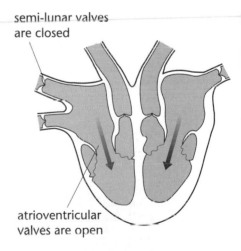

semi-lunar valves are closed

atrioventricular valves are open

Stage 2

Ventricular systole, atrial diastole
Both atria then relax. Both ventricles contract simultaneously. This results in higher pressure in the ventricles compared to the atria above. The difference in pressure closes each atrioventricular valve. This prevents backflow of blood into each atrium. Higher pressure in the ventricles compared to the aorta and pulmonary artery opens the semi-lunar valves and blood is ejected into these arteries. So blood flows through the systemic circulatory system via the aorta and vena cava and through the lungs via the pulmonary vessels.

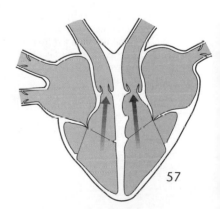

Stage 3

Ventricular diastole, atrial diastole

Immediately following ventricular systole, both ventricles and atria relax for a short time. Higher pressure in the aorta and pulmonary artery than in the ventricles closes the semi-lunar valves. This prevents the backflow of blood. Higher pressure in the vena cava and pulmonary vein than in the atria results in the refilling of the atria.

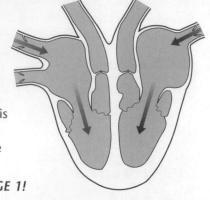

The cycle is now complete – *GO BACK TO STAGE 1!*

> Returning to Stage 1, the cycle begins again. The hormone adrenaline increases the heart rate still further. Even your examinations may increase your heart rate!

The whole sequence above is **one** cardiac cycle or heartbeat and it takes less than one second. The number of heartbeats per minute varies to suit the activity of an organism. Vigorous exercise is accompanied by an increase in heart rate to allow faster collection, supply and removal of substances because of enhanced blood flow. Conversely during sleep, at minimum metabolic rate, heart rate is correspondingly low because of minimum requirements by the cells.

How is the heart rate controlled?

> **SAN**
>
> **AVN**
>
> **Purkinje tissue**
>
> The SAN is the natural pacemaker of the heart.
>
> The electrical activity of the heart is shown in an ECG

It has already been stated that the cardiac muscle cells have their own inherent rhythm. Even an individual cardiac muscle cell will contract and relax on a microscope slide under suitable conditions. An orchestra would not be able to play music in a coordinated way without a conductor. The cardiac muscle cells must be similarly coordinated, by a **pacemaker** area in the heart. Electrical stimulation from the brain can alter the activity of the pacemaker and therefore change the rate and strength of the heartbeat.

- The heart control centre in the brain is in the medulla oblongata.
- The sympathetic nerve stimulates an increase in heart rate.
- The vagus nerve stimulates a decrease in heart rate.

> All of the Purkinje fibres together are known as the **Bundle of His**.

- These nerves link to a pacemaker structure in the wall of the right atrium, the **sinoatrial node (SAN)**.
- A wave of electrical excitation moves across both atria.
- They respond by contracting (the right one slightly before the left).
- The wave of electrical activity reaches a second pacemaker, the **atrioventricular node (AVN)**, which conducts the electrical activity through the **Purkinje fibres**.
- These Purkinje fibres pass through the septum of the heart deep into the walls of the left and right ventricles.
- The ventricle walls begin to contract from the apex (base) upwards.
- This ensures that blood is ejected efficiently from the ventricles.

> This is one of the examiners' favourite ways to test heart-related concepts. Look at the **peak** of the **ventricular contraction**. It coincides with the **trough** in the **ventricular volume**. This is not surprising, because as the ventricle contracts it empties! Use the data of higher pressure in one part and lower in another to explain:
>
> (a) movement of blood from one area to another
>
> (b) the closing of valves.

Graphs to show the changes in pressure and volume during the cardiac cycle

maximum contraction of ventricle

semi-lunar valve closed because aorta is at higher pressure than ventricle

atrium contracting

ventricle contracting

ventricle is lower in pressure than atrium so atrium fills ventricle

pressure/kPa

atrial systole   ventricular systole   diastole

volume of ventricle/cm³

volume of ventricle decreasing

volume of ventricle increasing

time/s

Progress check

The medulla oblongata can increase the heart rate. The statements below include all of the events which take place, but in the wrong order. Write them out in the correct sequence.

A this ensures that blood is ejected efficiently from the ventricles

B the wave of electrical activity reaches the **atrioventricular node (AVN)** which conducts the electrical activity through the **Purkinje fibres**

C a wave of electrical excitation moves across both atria

D the sympathetic nerve conducts electrical impulses

E electrical impulses are received at the **sinoatrial node (SAN)**

F as a result the atria contract

G the ventricle walls begin to contract from the apex (base) upwards

D E C F B C A

5.3 Blood vessels

After studying this section you should be able to:

- describe the structure and functions of arteries, capillaries and veins
- understand the importance of valves in the return of blood to the heart
- understand the difference between plasma, tissue fluid and lymph

LEARNING SUMMARY

Arteries, veins and capillaries

OCR 1.2.2

The blood is transported to the tissues via the vessels. The main propulsion is by the ventricular contractions. Blood leaves the heart via the arteries, reaches the tissues via the capillaries, then returns to the heart via the veins. Each blood vessel has a space through which the blood passes; this is the **lumen**. The structure of the vessels is shown below.

Artery

- It has a thick **tunica externa** which is an outer covering of tough collagen fibres.
- It has a **tunica media** which is a middle layer of **smooth muscle** and **elastic fibres**.
- It has a lining of **squamous endothelium** (very thin cells).
- It can contract using its **thick muscular layer**.

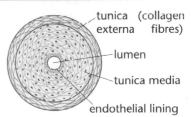

tunica (collagen
externa fibres)

lumen

tunica media

endothelial lining

Capillary

- It is a very thin blood vessel, the endothelium is just one cell thick.
- Substances can exchange easily.
- It has such a high resistance to blood flow that blood is slowed down. This gives more time for efficient exchange of chemicals at the tissues.

endothelium

lumen

Note that the pressure in the **arteries** is highest because:

(a) they are closest to the ventricles
(b) they contract forcefully themselves.

Capillaries are the next highest in pressure, the main factor being their resistance to blood flow.

Finally, the pressure of **veins** is the lowest because:

(a) they are furthest from the ventricles
(b) they have a low amount of muscle.

If given blood pressures of vessels, be ready to predict the correct direction of blood flow.

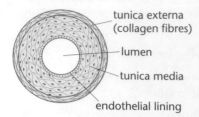

tunica externa (collagen fibres)

lumen

tunica media

endothelial lining

Vein

- It has a thin **tunica externa** which is an outer covering of tough collagen fibres.
- It has a very thin **tunica media** which is a middle layer of **smooth muscle** and **elastic fibres**.
- It has a lining of **squamous endothelium** (very thin cells).
- It is lined with semi-lunar valves which prevent the backflow of blood.

direction of blood flow

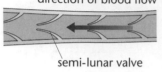

semi-lunar valve

How do the veins return the blood to the heart?

Veins have a thin tunica media, so only mild contractions are possible. They return blood in an unexpected way. Every time the organism moves physically, blood is squeezed between skeletal muscles and forced along the vein.

> Blood must travel towards the heart because of the direction of the semi-lunar valves. Any attempt at backflow and the semi-lunar valves shut tightly!

KEY POINT

Capillary network

Every living cell needs to be close to a **capillary**. The arteries transport blood from the heart but before entry into the capillaries it needs to pass through smaller vessels called **arterioles**. Many arterioles contain a ring of muscle known as a **pre-capillary sphincter**. When this is contracted the constriction shuts off blood flow to the capillaries, but when it is dilated, blood passes through. Some capillary networks have a shunt vessel. When the sphincter is constricted blood is diverted along the shunt vessel so the capillary network is by-passed. After the capillary network has permeated an organ the capillaries link into a **venule** which joins a **vein**.

In the skin the superficial capillaries have the arteriole/shunt vessel/venule arrangement as shown opposite. When the arteriole is dilated (**vasodilation**) more heat can be lost from the skin. When the arteriole is constricted (**vasoconstriction**) the blood cannot enter the capillary network so is diverted to the core of the body. Less heat is lost from the skin.

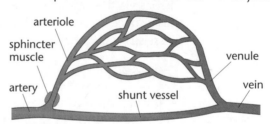

arteriole

sphincter muscle

venule

artery

shunt vessel

vein

Although the pressure of the blood in the capillaries is lower than in the arteries or arterioles, there is still enough pressure to force out some of the liquid part of the blood. The liquid part of the blood is called **plasma** and when it is forced out of the capillaries it is called **tissue fluid**.

This tissue fluid bathes the cells, supplying them with nutrients and taking up waste products. At the venous end of the capillary bed, most of this tissue fluid is reabsorbed back into the capillaries.

Lymphatic system

There is a network of vessels other than the blood system. They are the **lymphatic vessels**. They collect any tissue fluid that is not reabsorbed back into the capillaries.

The lymph vessels have valves to ensure transport is in one direction. Along some parts of the lymphatic system are lymph nodes. These are swellings lined with white blood cells (macrophages and lymphocyte cells). The lymph fluid is finally emptied back into the blood near the heart.

Lymph and tissue fluid do not contain red blood cells and usually contain less protein than the blood.

Progress check

The diagram shows the structure of a blood vessel.

(a) (i) Which type of vessel, artery, vein or capillary is shown? Give a reason for your choice.

(ii) What is the function of the tunica media?

(b) The pressure values 30 kPa, 10 kPa and 5 kPa correspond to the different types of vessel. Give the correct value for each vessel so that blood flows around the body.

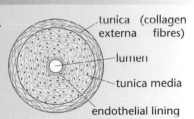

tunica (collagen externa fibres)

lumen

tunica media

endothelial lining

(a) (i) artery; the vessel has a thick tunica externa
(ii) contracts to help transport blood.
(b) artery, 30 kPa, capillary, 10 kPa and vein, 5 kPa.

5.4 The transport of substances in the blood

After studying this section you should be able to:

- describe the transport of oxygen in the blood and explain how oxygen is released at the tissues
- describe the transport of carbon dioxide

LEARNING SUMMARY

How is oxygen transported?

OCR ▷ 1.2.2

Oxygen is absorbed in the lungs from fresh air which has been breathed in. Red blood cells (**erythrocytes**) contain the protein **haemoglobin** which can reversibly combine with oxygen. In the lungs, where the concentration of oxygen is high, haemoglobin will take up oxygen and form oxyhaemoglobin. In the tissues where the oxygen concentration is low, the oxyhaemoglobin will dissociate and release the oxygen. This is shown by the graph below which is called the oxygen dissociation curve.

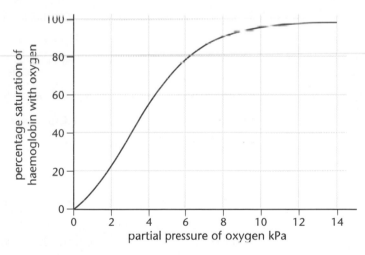

Features of the dissociation curve

- At high partial pressures of oxygen, haemoglobin has a high affinity (attraction) for oxygen and is highly saturated.
- At low partial pressures, the affinity is lower and the oxyhaemoglobin dissociates and is less saturated.
- The curve is sigmoid or 'S' shaped. This means that the curve is steep and a small change in partial pressure causes a massive loading or unloading of oxygen.

Changes to the dissociation curve

Different factors can cause changes to the dissociation curve:

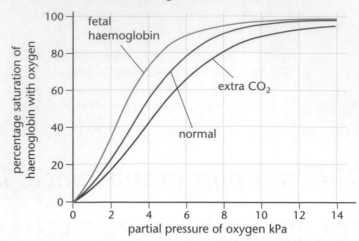

- The greater the amount of carbon dioxide at the tissues, the more the dissociation curve is moved to the right, and the more oxygen is 'off-loaded' to the tissues. This is called the Bohr shift. The carbon dioxide lowers the affinity of the haemoglobin for oxygen.

- Fetal haemoglobin has a greater affinity for oxygen than adult haemoglobin. This allows the fetus to take oxygen from the mother's haemoglobin.

Transport of carbon dioxide

OCR 1.2.2

This is done with the help of the red blood cells as follows:

- carbon dioxide diffuses into red blood cells from the tissues
- the carbon dioxide reacts with water to produce carbonic acid, this reaction being catalysed by the enzyme **carbonic anhydrase** in the cell (*a very fast reaction!*)

$$\overset{\text{carbonic anhydrase}}{H_2O + CO_2 \rightleftharpoons H_2CO_3}$$

water   carbon $\rightleftharpoons$ carbonic
dioxide   acid

- the carbonic acid **ionises** into H^+ and HCO_3^-

$$H_2CO_3 \rightleftharpoons H^+ + HCO_3^-$$

- haemoglobin combines with H^+ ions forming **haemoglobinic acid** which is very weak

$$H^+ + Hb \rightleftharpoons HHb$$

- HCO_3^- ions diffuse into the blood plasma to be transported to the lungs
- Cl^- ions diffuse into the red blood cell from the plasma; this counteracts the build up of positive charge from the H^+ ions. This is known as the **chloride shift**.

Do not confuse carbaminohaemoglobin with carboxyhaemoglobin which is formed when carbon monoxide combines with Hb.

The whole process is reversed once the blood reaches the lungs.

Most of the carbon dioxide in the blood is carried in this way as HCO_3^- ions. However, a small amount is carried combined with haemoglobin as carbaminohaemoglobin. Some is also dissolved in the plasma.

Plasma

OCR 1.2.2

This is the fluid in which all of the blood contents are transported. Listed below are some substances transported in the plasma:

- **water** – dissolves substances such as glucose for transport, stores dissolved prothrombin and fibrinogen which may be used later in clotting
- **proteins** – some are used to buffer the pH of the blood
- **glucose** – on its way to releasing energy in respiration
- **lipids** – on their way to releasing energy in respiration
- **amino acids** – on their way to cells to help assemble proteins or release energy in respiration
- **salts** – contribute to the water potential of blood, so that cells are not dehydrated by osmosis
- **hormones** – chemical messenger-molecules on their way to a target organ
- **antigens** – recognition proteins preventing white blood cells from destroying the person's own blood
- **antibodies** – made by lymphocytes to destroy antigens
- **urea** – made in the liver from excess amino acids, extracted by the kidneys.

Blood has a major role in the defence against disease (see the immune system page 100).

> Water has many important functions in the body, including being transported to the sweat glands to cool the body down.

> Note that the list outlines just some of the functions of plasma-transported substances. There are many more!

5.5 The transport of substances in a plant

After studying this section you should be able to:

LEARNING SUMMARY

- *recall the structure of a root and understand how water and ions are absorbed*
- *recall the structure of xylem and phloem and explain the processes by which they transport essential chemicals*
- *understand how plants lose water and how this loss can be measured*
- *recall the adaptations of xerophytes*

Root structure and functions

OCR 1.2.3

The roots of a green plant need to exchange substances with the soil environment. The piliferous zone just behind a root tip has many root hairs which have a high surface area to volume ratio.

- Root hairs are used for absorption of water and mineral ions and the excretion of carbon dioxide.
- They have a cell membrane with a high surface area to volume ratio to efficiently absorb water, mineral ions and oxygen, and excrete carbon dioxide.
- They project out into the soil particles which are surrounded by soil water at high water potential compared with the low water potential of the contents of the root hairs.
- They have a cell membrane which is partially permeable to allow water absorption by osmosis (see page 49).
- As they absorb more water by osmosis, the cell sap becomes more dilute compared with neighbouring cells. Water therefore moves to these adjacent cells which become more diluted themselves, so osmosis continues across the cortex.
- They have carrier proteins in the cell membranes to allow mineral ions to be absorbed by active transport.

> Note that the root hairs also absorb oxygen from the air to aid aerobic respiration. The high surface area to volume ratio certainly helps!

> Remember that water moves from a higher water potential to a more negative water potential.

> Remember that active transport needs energy, so mitochondria will be close to the carrier molecules on the membranes.

Passage of water into the vascular system

Once absorbed by osmosis, water needs to pass to the xylem vessels in order to move up the plant. First it must move across the cortex of the root and through the endodermis before entering the xylem. The mechanism of passage is not known but there are three theories:

- **apoplast** route, where the water is considered to pass between the cells
- **symplast** route, where the water is considered to pass via the cytoplasm of the cells via **plasmodesmata** (cytoplasmic strands connecting one cell to another)
- **vacuolar** route, where the water is considered to pass through the tonoplast then through the sap vacuole of each cell.

> Note the different theories for water transport across the width of the root.

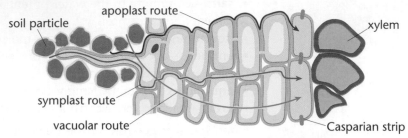

movement of water across the cortex

Casparian strip

Water moves across the cortex and needs to pass through the endodermal cells before entering the xylem vessels of the vascular system. Each cell of the endodermis has a waterproof band around it, just like a ribbon around a box. This means that water must pass through the cell in some way, rather than around the outside. If water moves by the apoplast route up to this point, then it must now move into the symplast or vacuolar pathways.

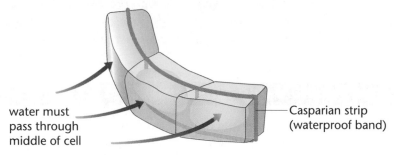

Casparian strip of the endodermal cells

Once the water has passed through the endodermis and navigated the pericycle then it must pass into the xylem for upward movement to the leaves and to the tissues.

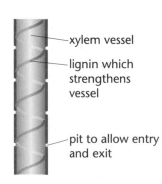

xylem vessel

lignin which strengthens vessel

pit to allow entry and exit

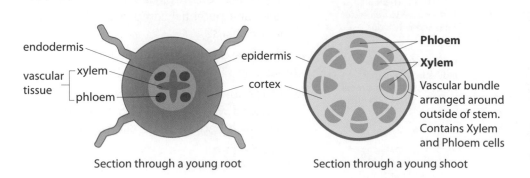

Section through a young root

Section through a young shoot

How does water move up the vascular system to the leaves?

Water moves into the **xylem** vessels in the vascular system in the centre of the root; it enters via **bordered pits**. The xylem is internally lined with **lignin**. This substance is waterproof and it also gives great strength to the xylem vessels, which are tube shaped. Much of the strength of a plant comes from cells toughened by lignin. A Giant Redwood tree is many metres high but water is still able to reach all the cells. Water moves up the xylem for the following reasons.

Remember that the xylem is part of the mass flow system ensuring that all cells receive their requirements.

- **Root pressure** gives an initial upward force to water in the xylem vessels. This can be shown by cutting off a shoot near soil level. Some sap will pour vertically out of the xylem of the remaining exposed xylem.

- Water moves up the xylem by **capillarity** which is the upward movement of a fluid in a narrow bore tube – xylem has very narrow vessels.

The factors in the list are known as the cohesion-tension theory and explain how water moves up the xylem.

- Capillarity occurs because the water molecules have an attraction for each other (cohesion) so when one water molecule moves others move with it

- Capillarity has another component – the fact that the water molecules are attracted to the sides of the vessels pulls the water upwards (adhesion).

- Transpiration causes a very negative water potential in the mesophyll of the leaves. Water in the xylem is of higher water potential and so moves up the xylem.

Xylem vessels die at the end of their maturation phase. The lignin produced inside the cells finally results in death. The young xylem cells end to end, finally produce a long tube-like structure (vessel) through which water passes. Xylem can still transport water after the death of the plant.

Mineral ions are also transported in the xylem.

Translocation

There is some uncertainty as to where the energy is required for translocation. Most theories involve active loading of sugars into the phloem at the source.

This is an **active process** by which **sugars** and **amino acids** are transported through the phloem. Sugar is produced in the photosynthetic tissues and must be exported from these **sources** to areas of need, i.e. usually areas which have large energy requirements. These areas are called sinks, e.g. terminal buds and roots.

> **KEY POINT**
>
> Roots cannot photosynthesise so they need carbohydrates to be supplied by other parts of the plant such as the leaf or storage organs.

The sugars are transported in the phloem which consists of two types of cell, the **sieve tube** and **companion cell**. Unlike xylem, the cells of the phloem are living.

Structure of the phloem tissue

The sieve tube has no nucleus so that essential proteins for life are made by the companion cell which does possess a nucleus. The companion cell maintains services to the sieve tube.

- Each **sieve tube** links to the next via a **sieve plate** which is perforated with pores.

- The sieve tube has cytoplasm and a few small mitochondria.

- Sugars are thought to pass through the sieve tubes by **cytoplasmic streaming**.

- The sieve tubes have no nucleus but are alive because of **cytoplasmic connections (plasmodesmata)** with the companion cell.

- Each companion cell has a nucleus and mitochondria.

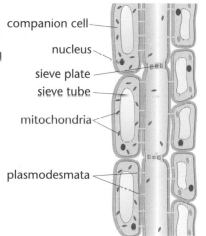

companion cell
nucleus
sieve plate
sieve tube
mitochondria
plasmodesmata

Radioactive labelling

This technique has been used to investigate the mechanism of translocation. It involves the use of a substance such as radioactive CO_2. The radioactive isotope, ^{14}C is used to make $^{14}CO_2$. A leaf is allowed to photosynthesise in the presence of $^{14}CO_2$ and makes the radioactive sugar ($^{14}C_6H_{12}O_6$). The route of the radioactive sugar can be traced using a Geiger-Müller counter. The greater the number of radioactive disintegrations per unit time, the greater the concentration of the sugar in that part of the plant.

How is water lost from a leaf?

OCR   1.2.3

Water moves up the xylem and into the mesophyll of a leaf. The process by which water is lost from any region of a plant is **transpiration**. Water can be lost from areas such as a stem, but most water is lost by **evaporation** through the **stomata**. Each stoma is a pore which can be open or closed and is bordered at either side by a guard cell. The diagrams show an open stoma and a closed stoma.

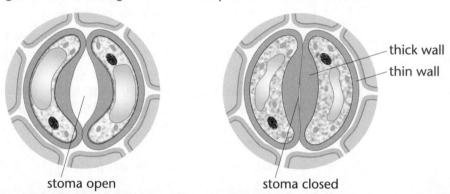

thick wall

thin wall

stoma open   stoma closed

Note that a living shoot may be photosynthesising whilst attached to the instrument. Only a minute amount of water would be used in this process. The instrument gives an accurate measure of transpiration.

KEY POINT

Transpiration from a leaf takes place as follows:
* the air spaces in the mesophyll become **saturated** with water vapour (**higher water potential**)
* the air outside the leaf may be of **lower humidity (more negative water potential)**
* this causes water molecules to **diffuse** from the mesophyll of the leaf to the outside.

Some water can escape through the cell junctions and membranes. This is known as **cuticular transpiration**. In the dark all stomata are closed. Even so, there is still water loss by cuticular transpiration.

How do the guard cells open and close?

In the presence of light:

* K^+ ions are actively transported into the guard cells from adjacent cells
* malate is produced from starch
* K^+ ions and malate accumulate in the guard cells
* this causes an influx of water molecules
* the cell wall of each guard cell is thin in one part and thick in another
* the increase in hydrostatic pressure leads to the opening of the stomata.

Closing of the stomata is the reverse of this process. Under different conditions the stomata can be partially open. The rate of transpiration can increase in warm, dry conditions or decrease at the opposite extreme.

Measuring the rate of transpiration

This is done indirectly by using a potometer. This instrument works on the following principle: for every molecule of water lost by transpiration, one is taken up by the shoot.

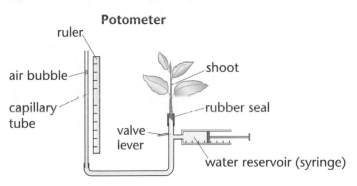

Potometer

Note that a living shoot may be photosynthesising whilst attached to the instrument. Only a minute amount of water would be used in this process. The instrument gives an accurate measure of transpiration.

The potometer is used as follows:

- a shoot is cut and the end is quickly put in water to prevent an air lock in the xylem
- the potometer is filled under water so that the capillary tube is full
- all air bubbles are removed from the water
- the shoot is put into the rubber seal
- the valve is changed to allow water uptake
- the amount of water taken up by the shoot per unit time is measured
- the shoot can be tested under various conditions.

Xerophytes

OCR 1.2.3

These are plants which have **special adaptations** to survive in drying, environmental conditions where many plants would become desiccated and die. The plants survive well because of a combination of the following features.

- thick cuticle to reduce evaporation
- reduced number of stomata
- smaller and fewer leaves to reduce surface area
- hairs on plant to reduce air turbulence
- protected stomata to prevent wind access
- aerodynamic shape to prevent full force of wind
- deep root network to absorb maximum water
- some store of water in modified structures, e.g. the stem of a cactus.

In an exam you may be given a photomicrograph of a xerophytic plant which you have not seen before. Look for *some* of the features covered in the bullet points opposite.

A cactus is a good example of a xerophyte. It makes excellent use of what little water there is available, and holds on to what it does manage to absorb really well. Its cuticle and epidermis are so thick that metabolic water released from the cells during night time respiration is retained for photosynthesis during the day. Nothing is wasted!

Progress check

Water is absorbed into a plant by the root hairs.

(a) The water potential of the root hair cells is more negative than in the soil water. Is this statement true or false?

(b) Describe:

   (i) the apoplast route across the cortex

   (ii) the symplast route across the cortex.

(a) true

(b) (i) water is considered to pass on the outside of the cell membrane

(ii) water passes through the cytoplasm of the cells through plasmodesmata.

Sample question and model answer

(a) The graph shows the oxygen dissociation curve for human haemoglobin.

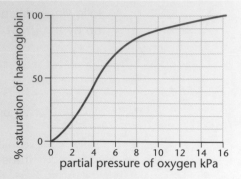

Note that haemoglobin is able to pick up a lot of oxygen, even at low partial pressure.

Use the information in the graph to help you answer the following questions.

(i) What is the advantage of haemoglobin as a respiratory pigment when oxygen in the air is at the low partial pressure 6 kPa? [1]

Even at a low partial pressure a lot of oxygen (70%) is taken up by the haemoglobin of a red blood cell.

The fact that haemoglobin is able to carry oxygen is important. However, it is just as important that the oxygen is off-loaded at tissues needing it. This is only possible because carbon dioxide is found at the tissues.

(ii) Explain the effect on the oxygen dissociation curve of a high partial pressure of carbon dioxide at a muscle. [2]

The curve is moved to the right and down so that oxygen is released.

(iii) Fetal haemoglobin has a greater affinity for oxygen than maternal haemoglobin. Draw a curve on the graph to show the oxygen dissociation curve for fetal haemoglobin. [1]

See graph opposite.

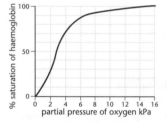

(b) The diagram shows **one** stage in the cardiac cycle.

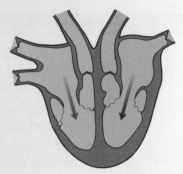

Always look for the valves. If the heart valve is open then the chamber behind it is contracting.

(i) Which stage of the cardiac cycle is shown in the diagram? Give **two** reasons for your answer. [3]

atrial systole
the atrioventricular valves are open/blood flows through the atrioventricular valves,
semi-lunar valves are closed.

(ii) Write an X in one chamber to show the position of the atrioventricular node (AVN). [1]

(iii) How does the AVN stimulate the contraction of the ventricles? [1]

Passes electrical impulses to Purkinje tissue/Bundle of His.

Practice examination questions

1 The diagram shows a capillary bed in the upper part of the skin. The arteriole is constricted.

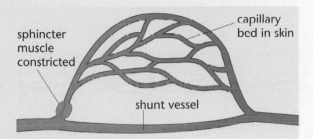

Use the information in the diagram and your own knowledge to answer the questions below.

(a) As a result of arteriole constriction, to where would the blood flow? [1]

(b) Explain how this would help maintain the body temperature. [4]

2 The table shows data about a person's heart before and after a training programme.

	Before training	After training
heart stroke rate	90 ml	120 ml
heart rate at rest	75 bpm	60 bpm
maximum heart rate	170 bpm	190 bpm

(a) Over a five minute period at rest before training, the cardiac output of the person was 33.75 litres.

How much blood would leave the heart, during the same time, whilst the person was at rest, after training? [2]

(b) After training, the maximum heart rate increased by 20 bpm. Explain the advantage of this increase to an athlete. [4]

(c) After training there are other changes in the body.

Explain:

(i) **two** changes which would improve the efficiency of the respiratory system. [2]

(ii) the effect of training on the muscles. [2]

Practice examination questions *(continued)*

3 The diagram shows a freshly cut, leafy shoot attached to a potometer. This was used to measure the amount of water taken up by the shoot under different conditions.

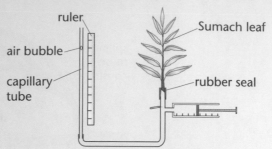

(a) What assumption must be made when using this apparatus to measure the rate of transpiration? [1]

(b) An air-lock can occur in the shoot which prevents water uptake.

    (i) In which plant tissue could an air-lock occur? [1]

    (ii) Describe the practical details by which a student could make sure that there was no air-lock in the shoot. [2]

(c) The radius of the capillary tube of the potometer was 1 mm. When a Sumach leaf was measured the air bubble moved 32 mm in one minute. Calculate the volume of water in mm³ which would be taken up by the leaf in one hour under the same environmental conditions. [3]

4 *Agave americana* is a xerophytic plant which grows in the deserts of Mexico.

Agave americana

Suggest **three** ways in which the plant is adapted to survive periods of very low rainfall. [3]

5 The diagram shows nerves linking the medulla oblongata with the heart.

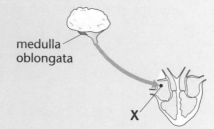

(a) Name part X. [1]

(b) What effect do the following have on the heart:

    (i) vagus nerve

    (ii) sympathetic nerve

    (iii) adrenaline? [3]

6 The diagram shows an aphid feeding on a plant. The sharp stylet is inserted into the phloem tissue which supplies the aphid with sucrose, plus organic and inorganic ions.

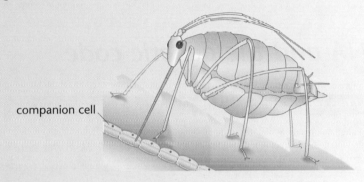

companion cell

(a)

(i) Name the phloem cell X from which the aphid obtains sucrose. [1]

(ii) Cell X does not have a nucleus or ribosomes, but still contains enzymes. Explain how this is possible. [3]

(b)

(i) Feeding aphids obtain the contents of the phloem without any sucking action being necessary. What does this indicate about the transport of substances through the phloem? [3]

(ii) Scientists investigated phloem contents by anaesthetising feeding aphids, then cutting their bodies from their stylets. Phloem contents oozed from the cut end of each stylet. The phloem contents were tested using iodine and heating with Benedict's solution.

	Tested with iodine	Heated with Benedict's solution
Contents of phloem	brown colour	brick-red colour

Referring to the results of the tests, explain what the scientists found out about the phloem contents using this method. [4]

(iii) Hot-wax ringing is a technique where hot wax is poured around a stem. This technique was used with the aphid method described in (ii). Radioactive carbon dioxide was supplied to one leaf so that a radioactive carbohydrate was made.

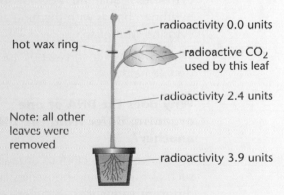

radioactivity 0.0 units

hot wax ring

radioactive CO_2 used by this leaf

radioactivity 2.4 units

Note: all other leaves were removed

radioactivity 3.9 units

Explain the effect of hot-wax ringing on the phloem tissue. [3]

Chapter 6

Genes and cell division

The following topics are covered in this chapter:

- DNA and the genetic code
- Cell division
- Gene technology

6.1 DNA and the genetic code

After studying this section you should be able to:

- recall the structure of DNA
- outline the roles of DNA and RNA in the synthesis of protein
- use organic base codes of DNA and RNA to identify amino acid sequences

LEARNING SUMMARY

Deoxyribonucleic acid (DNA) and chromosome structure

OCR 2.1.2

Each chromosome in a nucleus consists of a series of genes. A gene is a section of a chemical called **DNA** and each gene controls the production of a polypeptide important to the life of an organism. **Deoxyribonucleic acid (DNA)** is made up of a number of **nucleotides** joined together in a double helix shape.

You need to be aware that many nucleotides join together to form the polymer, DNA.

Each nucleotide consists of a phosphate group, a molecule of deoxyribose sugar and an organic base. Phosphate and pentose sugar units link to form the backbone of the DNA. Repeated linking of the monomer nucleotides forms the polynucleotide chains of DNA.

Each strand of DNA is said to be complementary to the other. **Examination tip:** be ready to identify one strand when given the matching complementary strand.

Each DNA molecule is made up of two polynucleotide chains. The two chains are held together by hydrogen bonds between the bases.

The organic base of each nucleotide can be any one of **adenine, thymine, cytosine** or **guanine**. Adenine forms hydrogen bonds with thymine and cytosine with guanine.

The two chains then twist up to form a double helix.

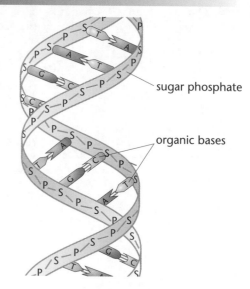

sugar phosphate

organic bases

A single nucleotide

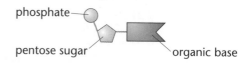

phosphate

pentose sugar

organic base

Why does the DNA of one organism differ from the DNA of another?

Differences in the DNA of organisms such as humans and houseflies lie in the **different sequences** of the organic bases. Each sequence of bases is a code to make a protein, usually vital to the life of an organism.

DNA

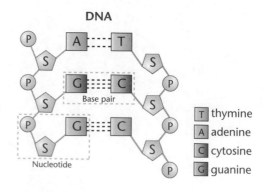

Base pair

Nucleotide

T thymine
A adenine
C cytosine
G guanine

How does DNA control protein production?

OCR ▸ 2.1.2

Each different polypeptide is made of a specific order of amino acids and so DNA must code for this order.

> Each three adjacent bases code for one amino acid in the polypeptide. The genetic code is therefore a **triplet** code.

Use the key to identify the amino acids in the table opposite.

Amino acid	Abbreviation
alanine	Ala
arginine	Arg
asparagine	Asn
aspartic acid	Asp
cysteine	Cys
glutamine	Gln
glutamic acid	Glu
glycine	Gly
histidine	His
isoleucine	Iso
leucine	Leu
lysine	Lys
methionine	Met
phenylalanine	Phe
proline	Pro
serine	Ser
threonine	Thr
tryptophan	Trp
tyrosine	Tyr
valine	Val

Do not learn all of the triplet codes. Be ready to use the supplied data in the examination. You will be given a key of different codes and functions.

If you are given a table of codes check them carefully. If the bases are from mRNA then there will be uracil in the table.

The table below shows all the triplet sequences of organic bases found along DNA strands and the coding function of each.

Genetic code functions of DNA

First	Second organic base								Third organic base
	A		**G**		**T**		**C**		
A	AAA	Phe	AGA	Ser	ATA	Tyr	ACA	Cys	A
	AAG		AGG		ATG		ACG		G
	AAT	Leu	AGT		ATT	stop	ACT	stop	T
	AAC		AGC		ATC	stop	ACC	Trp	C
G	GAA	Leu	GGA	Pro	GTA	His	GCA	Arg	A
	GAG		GGG		GTG		GCG		G
	GAT		GGT		GTT	Gln	GCT		T
	GAC		GGC		GTC		GCC		C
T	TAA	Ile	TGA	Thr	TTA	Asn	TCA	Ser	A
	TAG		TGG		TTG		TCG		G
	TAT		TGT		TTT	Lys	TCT	Arg	T
	TAC	Met	TGC		TTC		TCC		C
C	CAA	Val	CGA	Ala	CTA	Asp	CCA	Gly	A
	CAG		CGG		CTG		CCG		G
	CAT		CGT		CTT	Glu	CCT		T
	CAC		CGC		CTC		CCC		C

Each triplet code is **non-overlapping**. This means that each triplet of three bases is a code, then the next three, and so on along the DNA.

- AAA codes for the amino acid phenylalanine
- GAG codes for the amino acid leucine
- GAC codes for the amino acid leucine

There are more triplet codes than there are amino acids. This is known as the **degenerate code**, because an amino acid such as leucine can be coded for by up to six different codes. Some triplets do not code for amino acids but mark the beginning or end of polypeptides. They are **stop or start** triplets.

RNA and protein synthesis

DNA is found in the nucleus but proteins are made on ribosomes in the cytoplasm. Therefore a messenger is needed to transfer the code. This messenger is a molecule called ribonucleic acid (**RNA**).

RNA is a nucleic acid, made up of nucleotides like DNA but it has some important differences:

DNA	RNA
Two polynucleotide strands	One polynucleotide strand
Contains adenine, cytosine, guanine and thymine	Contains adenine, cytosine, guanine and the base uracil instead of thymine
Contains deoxyribose sugar	Contains ribose sugar

Messenger RNA is formed in the nucleus by making a complementary copy of the DNA coding for the polypeptide.

Here is an example of a coding strand of DNA:

DNA   A A A G A G G A C A C T   *(coding strand)*
mRNA   U U U C U C C U G U G A   *(messenger RNA)*

guanine (G) on DNA codes for cytosine (C) on mRNA

cytosine (C) on DNA codes for guanine (G) on mRNA

thymine (T) on DNA codes for adenine (A) on mRNA

adenine (A) on DNA codes for uracil (U) on mRNA

> This type of RNA is called messenger RNA (mRNA) because it carries the message out of the nucleus. There are two other types of RNA called rRNA and tRNA.

Progress check

(a) Name the parts of a nucleotide.
(b) (i) By which bonds do the two strands of DNA link together?
     (ii) How would these bonds be broken in the laboratory to produce single strands of the DNA?
(c) Which organic base is found in DNA but not in RNA?

(a) pentose sugar, phosphate and organic base. The organic base may be thymine, adenine, cytosine or guanine
(b) (i) hydrogen bonds (ii) heat
(c) thymine

6.2 Cell division

After studying this section you should be able to:

- describe and explain the semi-conservative replication of DNA
- understand that DNA must replicate before cell division can begin
- recall the purpose of mitosis and meiosis
- recognise each stage of the cell cycle and cell division by mitosis

LEARNING SUMMARY

How do cells prepare for division?

OCR   2.1.2

Before cells divide they must first make an exact copy of their DNA by using a supply of organic bases, pentose sugar molecules and phosphates. The method by which DNA is copied is called **semi-conservative replication**. The diagram (right) shows this taking place.

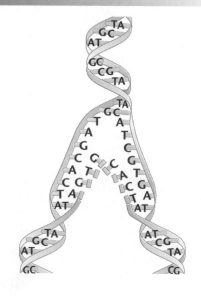

- The DNA begins to unwind under the influence of the enzyme DNA helicase.
- Hydrogen bonds between the two chains then break and the two strands separate.
- Each complementary strand then acts as a template to build its opposite strand from free nucleotides.
- The enzyme DNA polymerase joins the nucleotides together. This process results in the production of two identical copies of double-stranded DNA.

> Remember that as the DNA unwinds each single strand is a **complement** to the other. This means that each has the **matching** series of organic bases.

Evidence for the semi-conservative replication of DNA

The first real evidence came from the results of an experiment carried out by two researchers, Meselson and Stahl.

Bacteria were cultured with a heavy isotope of nitrogen located in the organic bases of their DNA.

The bacteria were then supplied with bases containing the normal light nitrogen atoms. They replicated their DNA using these bases. Their population increased.

Each molecule of DNA of the next generation had one strand containing heavy nitrogen and one strand containing light nitrogen. The mass of the DNA was therefore midway between the original heavy form and normal light DNA.

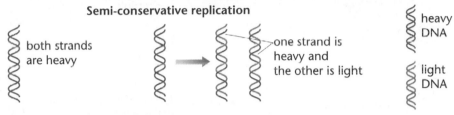

Semi-conservative replication

both strands are heavy → one strand is heavy and the other is light — heavy DNA, light DNA

Semi-conservative therefore means that as DNA splits into its two single strands, each of the new strands is made of newly acquired bases. The other strand, part of the original DNA, remains.

Types of cell division

OCR 1.1.3

Cells divide for the purposes of growth, repair and reproduction. Not all cells can divide but there are two ways in which division may occur: **mitosis** and **meiosis**.

In most organisms, the chromosomes in each cell can be arranged in pairs. Each cell therefore has two copies of each gene.

When the chromosomes are in pairs, the cell is said to be **diploid** and the pairs are called **homologous pairs**.

Mitosis produces genetically identical copies of cells with the same number of chromosomes for growth, repair and asexual reproduction.

Meiosis produces cells that have half the number of chromosomes, one from each pair. These cells are **haploid** and are used as gametes.

Meiosis introduces variation because the pairs of chromosomes can be split up in many different ways. This is called independent assortment.

The cell cycle and mitosis

OCR 1.1.3

The length of time between a cell being formed and it dividing is called the **cell cycle**. This can be divided up into a number of different phases:

- G1 phase – the cell grows making new proteins and more organelles
- S phase – the DNA of the chromosomes is replicated by semi-conservative replication
- G2 – more organelles are made and a spindle forms
- M phase – this is mitosis involving the separating of the genetic material into two nuclei
- C phase – cytokinesis, where the cell divides into two.

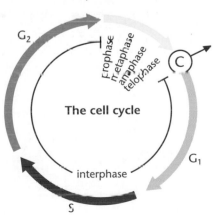

The cell cycle

Remember that DNA replication takes place before cell division in **interphase** (see page 75). This is not an integral phase of mitosis or meiosis.

The length of the cell cycle varies between different types of cells. Some cells never divide once formed but cells of the bone marrow divide about every eight hours. Interphase usually takes up 95% or more of the whole cell cycle.

Be ready to analyse photomicrographs of all phases of mitosis. If you can spot 10 pairs of chromosomes at the end of telophase, then this is the original diploid number of the parent cell.

Interphase is the period of time between cell divisions. It is made up of G1, S and G2 phases. Once cells start mitosis, they all go through a similar sequence of events. This is shown in the diagrams.

1 Prophase

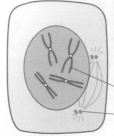

Each chromosome forms two chromatids joined by a centromere. Two centrioles begin to move forming a spindle.

chromatid

centriole

2 Metaphase

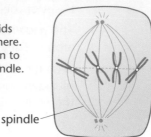

The chromatids, still joined by a centromere move to the middle of the cell. Each of the two chromatids has identical DNA to the other.

spindle

3 Anaphase

The spindle fibres join to the centromeres. The spindle fibres shorten and the centromeres split. The separated chromatids are now chromosomes.

4 Telophase

Identical chromosomes move to each pole. The nuclear membrane re-forms. The cell membrane narrows at the middle and two daughter cells are formed.

6.3 Gene technology

After studying this section you should be able to:

- *define different types of stem cells*

LEARNING SUMMARY

Stem cells

OCR ▶ 1.1.3

When cells have differentiated and become specialised they lose their ability to divide to form other types of cells. They also can only divide a limited number of times. **Stem cells** are undifferentiated cells that can divide to form different types of cells. There are different types of stem cells:

- Embryonic stem cells are **totipotent**. This means that they can form any type of cell.
- Later in the embryo and in the adult, the stem cells are **pluripotent**. This means that they can form certain types of cells.

Stem cells have the potential to be used in many types of medical therapies but the use of embryonic stem cells, in particular, has raised a number of ethical issues.

Sample questions and model answers

Always remember to read all information given. You must be aware of what the examiner is testing. Try to link the question with the part of the specification the topic comes from.

You will not be expected to remember all the genetic code! In this question you are given data. Make sure you have revised all key words so that you can apply the principles you have learned to the data.

The exam board never gives the full name of amino acids. There will always be a key.

The table below shows some mRNA codons and the amino acids which are coded by them.

	second position				
	U	**C**	**A**	**G**	
first position **U**	Phe	Ser	Tyr	Cys	U
	Phe	Ser	Tyr	Cys	C
	Leu	Ser	stop	stop	A
	Leu	Ser	stop	Trp	G

(first column = first position **U**; far right column = third position, with U, C, A, G)

Key to amino acids

Ser – serine   Tyr – tyrosine

Phe – phenylalanine   Trp – tryptophan

Leu – leucine   Cys – cysteine

Use the information in the table to help you answer the following questions.

1

There are two codes to choose from.

(a) Give a sequence of mRNA bases which would code for leucine.   [1]

UUA or UUG

(b) What does the mRNA base sequence UAC code for?   [1]

Tyrosine

Follow the table headings to indicate U (first base), A (second base) and C (third base).

This can be tricky! Remember that you need to work backwards. Given mRNA you know that it is coded for by one strand of DNA. Work out one DNA strand then use T – A, and G – C links.

2

The mRNA sequence UCA codes for serine. Work out the base pairs on the DNA. [3]

UCA is coded for by these bases: AGT

AGT links to the bases TCA

So the DNA is AGT
              TCA

If you failed to learn the definition of the degenerate code then you would fail to apply it to this data.

3

Use evidence from the table to show that serine is an example of the degenerate code.   [1]

It is coded for by four different base sequences.

4

UAG codes for 'stop'. Explain the effect of the 'stop' code during the process of protein synthesis.   [2]

It is responsible for the polypeptide being terminated which allows it to leave the ribosome once all the amino acids have been linked.

Practice examination questions

1 The diagram below shows a stage in the process of mitosis.

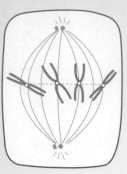

(a) Give the stage of mitosis shown. [1]

(b) How many chromosomes would there be in the daughter cells? [2]

2 The table below shows the relative organic base proportions found in human, sheep, salmon and wheat DNA.

Organism	Proportion of organic bases in DNA (%)			
	Adenine	Guanine	Thymine	Cytosine
human	30.9	19.9	29.4	19.8
sheep	29.3	21.4	28.3	21.0
salmon	29.7	20.8	29.1	20.4
wheat	27.3	22.7	27.1	22.8

(a) Refer to the proportion of organic bases in salmon DNA to explain the association between specific bases. [2]

(b) Suggest a reason for the small difference in proportion of the organic bases adenine and thymine in sheep. [1]

(c) All species possess adenine, guanine, thymine and cytosine in their DNA. Account for the fact that each species is different. [2]

Chapter 7

Classification and biodiversity

The following topics are covered in this chapter:

- Classification
- Biodiversity

- Maintaining biodiversity

7.1 Classification

After studying this section you should be able to:

- understand why and how organisms are classified
- describe the binomial system of naming organisms
- describe some of the more recent techniques used to classify organisms

Classifying and naming

OCR   2.3.2

It is estimated that there are more than 10 million different kinds of living organisms alive on Earth. The study of the range of different organisms is called **systematics**. For centuries, scientists have tried to classify organisms into groups and give them names.

> **KEY POINT**
>
> The system of classifying organisms into groups is called **taxonomy** and the system of naming is called **nomenclature**.

Organisms are put into groups based on the largest number of common characteristics. Firstly, they are put into one of five large groups called kingdoms.

The table shows the characteristics of organisms in the five kingdoms:

Try this!

Pretty **P**olly **F**inds **P**arrots **A**ttractive.

It will help you to remember the kingdoms!

You will probably be tested on your knowledge of characteristics across the kingdoms.

Information will usually be given for any subgroup tasks.

The table includes some of the main features of each kingdom.

Prokaryotae and Protoctista tend to give more of a challenge than some of the other kingdoms, and are examined more often.

→ increase in complexity

Prokaryotae	Protoctista	Fungi	Plantae	Animalia
very simple cells with few organelles	unicellular cells with membrane-bound organelles	heterotrophic nutrition	multicellular organisms which are photosynthetic	multicellular organisms which are heterotrophic
no membrane-bound organelles	some are photosynthetic, but many have heterotrophic nutrition	some saprotrophic, some parasitic	cells have cellulose cell wall, sap vacuole and chloroplasts	no cell walls, no sap vacuoles
if there are flagellae, then not 9 + 2 system of microtubules		consists of thread-like hyphae, chitin cell walls		
DNA in strands, no true nucleus	reproduction usually involves fission	many nuclei in hyphae, not in one per cell organisation	reproduce by seeds, or by spores, some sexual, some asexual.	
e.g. bacteria and cyanobacteria	e.g. algae and protozoa	reproduction involves the production of spores		

Each kingdom can be subdivided into a number of progressively smaller groups. Ultimately, this leads to an individual type of organism, a **species**.

The seven groups can be difficult to remember. Try the easy way!

King Penguins Climb Over Frozen Grassy Slopes

The first letter of each word will help you remember. It is a mnemonic which is an excellent strategy to aid recall.

The hierarchy of the groups is shown below.

	Example 1	Example 2
Kingdom	Animalia	Animalia
Phylum	Chordata	Arthropoda
Class	Mammalia	Insecta
Order	Primates	Lepidoptera
Family	Hominidae	Pieridae
Genus	Homo	Pieris
Species	sapiens	brassica

What is a species?

The smallest classification group is the species.

> Organisms are members of the same species if they can breed together to produce **fertile offspring**.

KEY POINT

Remember, similar organisms such as horses and donkeys can mate to produce mules, but mules are infertile.

Naming organisms

Most organisms are known by common names. We use these names all the time. The problem is that an organism may be known by different names or sometimes different organisms can have the same common name. A scientist called Linnaeus introduced a scientific naming system so that each species could have a unique scientific name. This system is called the **binomial system**.

The name consists of two parts, the name of the genus written with a capital letter and the name of the species, written in lower case.

Notice that binomial names are always typed in italics or underlined if handwritten.

Panthera leo

Panthera tigris

Modern classification techniques

OCR 23.2

For centuries, organisms were classified according to observable physical features. This might be the structure of an animal's skull or the shape of a plant's leaves. Scientists can now use a range of different techniques.

- Microscopic structure – modern electron microscopes have shown that bacteria and other single-celled organisms such as amoeba have completely different cell structure and so they are now put into different kingdoms: *Prokaryotae* and *Protoctista*.

- Genetic differences – it is now possible to compare the DNA base sequences of different organisms. This can be done by a process called **DNA hybridisation**. Short, single-stranded sections of DNA are produced from one species and the extent to which they bind with DNA from another species is measured.
- Biochemical differences – the occurrence of different biochemical molecules is often a good indicator of relationships. It is also possible to work out and compare the amino acid sequence of common proteins.
- Immunological evidence – antibodies against human proteins can be made and then tested against proteins from other animals. The more effective the antibodies are, then the closer the evolutionary relationship is between humans and the other animal.

The aim of modern classification systems is to use a range of techniques to produce a system that classifies organisms based on their **evolutionary relationships**.

> **KEY POINT**
> A system based on evolutionary relationships is called a **phylogenetic system**.

Immunological results give the following % similarities to man:

chimpanzee 97%,
gibbon 92%,
lemur 37%,
pig 8%

7.2 Biodiversity

After studying this section you should be able to:

LEARNING SUMMARY

- *understand what is meant by the term biodiversity*
- *explain why there is so much biodiversity on Earth*

What is biodiversity?

OCR 2.3.1

Biodiversity is a measure of the variation between different living organisms. It could be:
- **genetic diversity** – the differences between the genes in a species
- **species diversity** – the number of different species in a community
- **ecosystem or habitat diversity** – the variety of different areas where organisms can live.

A few definitions

A **habitat**: an area in which an organism lives.
A **population**: all the organisms of one species living in a habitat.
A **community**: all the different species (different populations) living in a habitat.
An **ecosystem**: all the living organisms (biotic) and inorganic parts (abiotic) of a habitat.
A **niche** is where an organism lives and its role in that ecosystem.

A pond is a habitat containing populations such as stickleback and pondweed. All the organisms make up the community and the organisms, the water, the mud, etc. are an ecosystem.

Estimating populations

Often a full count of organisms in an ecosystem is not possible because of the size of the ecosystem. A **sampling technique** is used which requires a **quadrat**. This is a small area enclosed by wire or wood, around 0.25 m². When placed in the ecosystem the organisms inside the area can be counted, as well as the **abiotic factors** which influence their distribution. Ecologists use units to measure organisms within the quadrats. Frequency (f) is an indication of the presence of an organism in a quadrat area. This gives no measure of numbers. However the usual unit is that of density, the numbers of the organism per unit area. Sometimes percentage cover is used, an indication of how much of the quadrat area is occupied.

Point quadrat

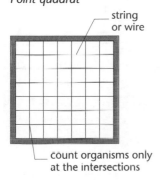

string or wire

count organisms only at the intersections

Consider a survey of two species *Taraxacum officinale* (dandelion) and *Plantago major* (Great plantain) of the lawn habitat shown below.

Lawn habitat

A simplified results table

Quadrat no.	Dandelion
1	2
2	12
3	15
4	3
5	4
6	8
7	7
8	10
9	9
10	15

mean = 8.5 per quadrat
Dm^{-2} = 34

dandelion

great plantain

Make sure you can use a key to identify different organisms. An example is given in the practice questions on page 88.

It is important to use a suitable technique when surveying with quadrats. When you observe a habitat which appears **homogeneous** or **uniform**, like a field yellow with buttercups, then you should use **random quadrat placement**. The area should be gridded, numbers given to each sector of the grid, then a random number generator used. The probability of the numbers in one quadrat representing a field would be very low. In practice the **mean** numbers from large numbers of quadrats do represent the true numbers in a habitat.

Simpson's Index of Diversity

This is used as a measure of the range and numbers of species in an area.

$$D = 1 - \Sigma \left(\frac{n}{N}\right)^2$$

N = total no. of all individuals of all species in the area
n = total no. of individuals of one species in an area
Σ = the sum of

Consider this example of animals in a small pond

In another pond there were:

crested newt	45
stickleback	4
leech	18
great pond snail	10

D = 0.58

Look at both indices. 0.83 is an indicator of greater diversity. The higher number indicates greater diversity.

	n	$\frac{n}{N}$	$\left(\frac{n}{N}\right)^2$
crested newt	8	0.072	0.0052
stickleback	20	0.18	0.032
leech	15	0.14	0.020
great pond snail	20	0.18	0.032
dragonfly larva	2	0.018	0.00032
stonefly larva	10	0.09	0.008
water boatman	6	0.054	0.0029
caddisfly larva	30	0.27	0.073
	N = 111		0.17

D = 1 – 0.17
= 0.83

Sources of biodiversity

OCR 2.3.3

Remember that a species shows continuous variation when there are small incremental differences, e.g. height of people in a town. Beginning with the smallest and ending with the tallest there would probably be at least one person at each height, at 1cm increments. A smooth gradation of differences!

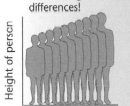

What is variation?

Species throughout the biosphere differ from each other.

Variation describes the differences which exist in organisms throughout the biosphere. This variation consists of differences **between** species as well as differences **within** the same species. Each individual is influenced by the environment, so this is another source of variation.

genotype + environment = phenotype

The alleles which are expressed in the phenotype can only perform their function efficiently if they have a supply of suitable substances and have appropriate conditions. Ultimately new genes and alleles have appeared by mutation. The spontaneous appearance of **advantageous** new mutations is also possible. This may lead to formation of new species.

Continuous variation

This is shown when there is a range of **small incremental differences** in a feature of organisms in a population. An example of this is height in humans. If the height of each pupil in a school is measured then from the shortest pupil to the tallest, there are very small differences across the distribution. This is shown by the graph below which shows smooth changes in height across a population. This type of variation is shown when features are controlled by **polygenic inheritance**. A number of genes **interact** to produce the expressed feature.

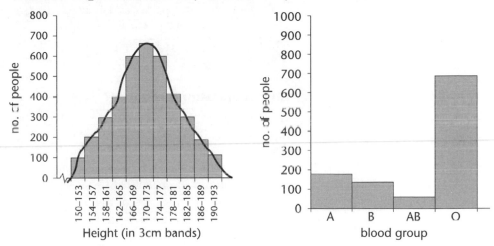

Discontinuous variation

This is shown when a characteristic is expressed in discrete categories. Humans have four discrete blood groups, A, B, AB or O. There are no intermediates, the differences are clear cut!

Mean and standard deviation

The variation shown by most populations can be described graphically. The variable usually forms a bell-shaped curve called a **normal distribution**.

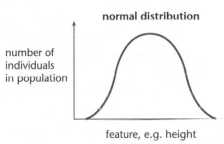

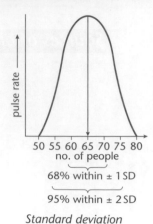

Standard deviation

In a normal distribution, 68% of the values are within one standard deviation either side of the mean and 95% are within two.

The **mean** is the value at the peak of the curve. The **standard deviation** describes how the values are spread out. The higher the standard deviation, then the more variation there is.

Why do organisms show variation?

The many different habitats on Earth provide different conditions for organisms to live in. There may be variations in a number of different factors that can affect organisms:

- **climatic** factors, e.g. temperature, water availability, light
- **edaphic** (soil) factors, e.g. pH, minerals and oxygen content
- **biotic** factors, e.g. competition.

Organisms have become adapted to living in a particular habitat with a particular set of factors. These factors may be behavioural, physiological and anatomical. The table below shows some examples.

Type of organism	Adaptation	Behavioural, physiological or anatomical
cacti	leaves are spines to reduce water loss	anatomical
birds	parental care shown to young	behavioural
carnivores and herbivores	particular dentitions adapted to deal with specialised diets	anatomical
llamas	production of haemoglobin molecules with a higher affinity for oxygen	physiological

How is variation produced?

The vast biodiversity of organisms on Earth has been produced by the process of **natural selection** over millions of years. These are the key features of natural selection:

The survival of the best adapted organisms is often called 'survival of the fittest'.

Populations of organisms show variation and some of this variation can be inherited.

More organisms are born than can survive in a particular environment.

There must be a struggle for survival. Those organisms that are best adapted to the environment will survive.

The organisms that survive will breed and pass on their genes.

The theory of natural selection was first put forward by Charles Darwin in 1858 in his book *On the Origin of Species*. The small changes to populations that occur over long periods of time could result in the formation of new species. This is called **speciation**.

Recently, natural selection has been used to explain the development of:

- pesticide resistance in insects
- drug resistance in microorganisms.

7.3 Maintaining biodiversity

After studying this section you should be able to:

- appreciate why it is important to maintain biodiversity
- describe some of the approaches used to try and maintain biodiversity

Why is it important to maintain biodiversity?

OCR 2.3.4

The human population of the Earth is rapidly increasing. This has resulted in increasing pollution and overexploitation of the Earth's natural resources. This has tended to reduce the biodiversity of many ecosystems and threaten entire habitats.

Species conservation involves managing the Earth's resources so that the biodiversity of animals and plants can be maintained. There are a number of reasons why people think that this is necessary.

- **Economic** – animal or plant species that are important to the economy of countries for food or raw materials may be lost.
- **Ecological** – the loss of species may have implications for the survival of other organisms.
- **Ethical** – many people think that we have no right to cause the extinction of other species.
- **Aesthetic** – many people like to enjoy the variety of organisms and habitats of the Earth.

Approaches to maintaining biodiversity

Conservation does not involve just preserving wildlife but requires active maintenance to counteract the damage being done by mankind. Many of the threats to habitats are changes that are occurring on a world-wide level. An example of this is global warming which may alter the climate of the whole planet. Successful conservation therefore requires action at **international, national** and **local** levels.

International conservation

It is often difficult but essential to get international agreement for conservation projects.

CITES

This is The Convention in International Trade in Endangered Species. It is an agreement between 172 different governments that restricts the trade of certain species. Lists of endangered species have been drawn up and roughly 5000 species of animals and 28 000 species of plants are protected by CITES against over-exploitation through international trade.

Rio Convention on Biodiversity

The ongoing arguments over whaling restrictions are a good example of how difficult it is to get binding international agreements.

This is an international treaty that was adopted at the Earth Summit in Rio de Janeiro in June 1992. The Convention has three main objectives.
1. The conservation of biological diversity (or biodiversity).
2. The sustainable use of biological resources.
3. The fair and equitable sharing of benefits arising from genetic resources.

National projects

Many countries have zoos or botanical gardens that have been popular with visitors for many years. Now many of these establishments are focusing more on the conservation of species rather than just entertaining the public. This has been encouraged by grants and charitable status.

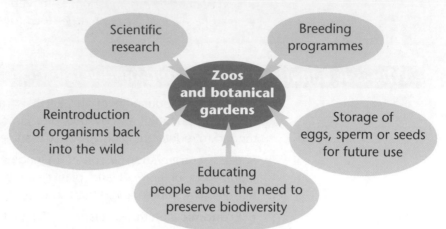

Although these zoos and botanical gardens are funded on a national basis, they still require international cooperation in terms of setting up breeding programmes and storing material. This is because it is vital to maintain genetic variety in order to prevent inbreeding.

Local conservation

It is now necessary for developers planning certain building projects to construct an Environmental Impact Statement before they ask for planning permission. The statement describes the likely significant effects of the development on the environment and biodiversity. The statement must be taken into account by the local planning authority before it may grant consent.

Sample question and model answer

The quagga is an extinct mammal that looked like a zebra but had stripes only on its head, neck and forebody. The quagga lived in Africa until the last animal was shot in the late 1870s. The last specimen died in 1883 in a zoo.

The quagga, named *Equus quagga*, was originally thought to be a separate species to the plains Zebra.

Over the last fifty years or so, the markings of many individual zebras have been recorded by scientists.

Because of the great variation in stripe patterns, taxonomists were left with a problem. Was the quagga a separate species or was it a type of zebra? The quagga was the first extinct creature to have its DNA studied. Recent research using DNA from preserved specimens has demonstrated that the quagga was not a separate species at all, but a variety of the plains zebra, *Equus burchelli*.

> Questions about natural selection always have similar mark schemes. Just make sure that you apply this mark scheme to the particular situation.

(a) What is the name of the genus that both the plains zebra and the quagga belong to? [1]

Equus

(b) What type of variation does the stripe pattern in zebras demonstrate? [1]

Continuous variation

(c) How could scientists use the DNA to investigate the relationship between the zebra and the quagga? [?]

Use DNA hybridisation.
The DNA is split into single strands and then scientists see how well it binds with zebra DNA strands.

(d) The plains zebra may have more stripes than the quagga because the stripes provide better camouflage on the plains.
How might this pattern have developed by natural selection? [3]

Zebras show variation and are born with different patterns of stripes.
More stripes make the zebra better camouflaged and so more likely to survive.
These zebras survive, breed and pass on the genes for more stripes.

Practice examination questions

1   The following key distinguishes between the five kingdoms.

1	Organisms without membrane bound organelles	A
	Organisms with membrane bound organelles	GO TO 2
2	Organisms have hyphae	B
	Organisms do not have hyphae	GO TO 3
3	Organisms unicellular or colonial	C
	Organisms not unicellular or colonial	GO TO 4
4	Organisms multicellular and have thylakoid membranes in some cells	D
	Organisms multicellular and have no thylakoid membranes in any cells	E

Name kingdoms A, B, C, D and E [5]

2   (a)   The song-thrush (*Turdus ericetorum*) and mistle-thrush (*Turdus viscivorus*) are in the same family, *Turdidae*. Large sections of their DNA are common to both species. Complete the table to classify both organisms. [3]

	Mistle-thrush	Song-thrush
Kingdom		
Phylum	Chordata	Chordata
	Aves	
	Passeriformes	Passeriformes
Genus		
Species		

[1]   [1]   [1]

(b)   How is it possible to find out if two female animals are from the same species? [1]

Practice examination questions

3   Students wished to investigate whether a pond in a field was being affected by a nearby cattle feeding station.
They sampled the water in the pond using a net and caught the following organisms:

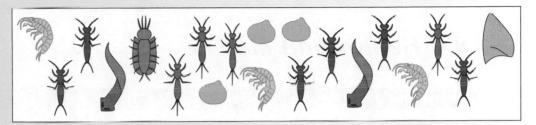

(a) Use the student' results to produce an index of diversity for the sample.   [2]

(b) The students sampled another pond in the same way.
This pond was further from the feeding station and their results produced an index of diversity of 0.85.
What conclusions can be drawn from the students' results?   [2]

Chapter 8

Human health and disease

The following topics are covered in this chapter:

- *Health and lifestyle*
- *Disease*

- *Immunity*
- *Assisting immunity*

8.1 Health and lifestyle

After studying this section you should be able to:

LEARNING SUMMARY

- *define health*
- *understand the link between diet and good health*
- *describe ways in which food technology is used to increase food supplies*
- *understand how coronary heart disease can be affected by lifestyle*
- *describe the specific effects of smoking tobacco*

How can we achieve good health?

OCR   2.2.2

Good health is not just an absence of disease or infirmity. It is the physical, mental and social well-being of a person. The development of a healthy person begins in the uterus. It is important that the mother supplies the fetus with suitable nutrients for development, e.g. amino acids for proteins essential for healthy growth. The mother's diet is important for both her and the fetus.

Following birth, the emotional and social development are equally as important as physical development.

If a person is healthy then they may expect the following:

- an absence of disease
- an absence of pain
- to be fit and have good muscle tone

- an absence of stress
- to get along with other people in society
- to have a long life expectancy.

The importance of diet

OCR   2.2.1

The human diet is vital to good health. A newly born baby needs to feed on mother's first milk (**colostrum**) which is rich in antibodies. This gives immunity against some diseases. It is important that each of the following food classes is included in a person's diet, in a suitable proportion.

Examiner's tip
It is likely that you will be supplied with data about dietary constituents. Be ready to analyse the data and apply the principles of a balanced diet. Try to remember the main function of each substance. Analyse the dietary reference values for food energy and nutrients in the UK. The values indicate amounts of individual food components required per day and those which should not be exceeded.

The balanced diet

- **Carbohydrates** – sugars and starch supply metabolic energy; cellulose (dietary fibre) stimulates peristalsis so that constipation is prevented.
  Source – potato and bread

- **Proteins** – supply metabolic energy and are needed in growth and repair. **All enzymes are proteins**. Very important!
  Source – meat and nuts

- **Fats and oils** – supply metabolic energy and are needed in cell membrane formation, as they help to make phospholipids.
  Source – butter and cooking oil

- **Vitamins** – organic substances needed in minute quantities to maintain health, e.g. vitamin A. This is essential to make the pigment in the rods of the retina.
  Source (vitamin A) – butter and carrots

- **Minerals** – inorganic ions needed for a number of important roles in the body, e.g. iron. This is essential for the production of haemoglobin and so is vital for oxygen transport.
  Source (iron) – red meat, spinach
- **Water** – makes up over 50% of the content of blood plasma. It is needed for many functions including as a solvent and cooling the body down.

If a person does not eat enough of any one constituent of their diet then there is a deficiency disease, e.g. a protein deficiency causes kwashiorkor. If a person eats too much carbohydrate and fats then obesity and cardiovascular problems can result. A balanced diet is vital!

> Consumption of an unbalanced diet can lead to malnutrition.

Remember that carbohydrates, proteins, lipids, vitamins, minerals, water and dietary fibre are all essential.

Daily energy requirement

Eating, then respiring carbohydrates, proteins, fats and oils, supplies the energy needed for good health. Energy content of food is usually measured in **kilojoules (kJ)**. A diet rich in carbohydrates and/or fats and oils which exceeds daily requirements results in **obesity**. Large quantities of fat are stored around the body resulting in cardiovascular problems. If the kilojoule intake is regularly less than the daily requirement then a condition known as **anorexia nervosa** can develop. People with this condition have an emotional disorder of which absence of appetite is a symptom.

Maintaining food supply

OCR 2.2.1

With the world's population continually growing, it is becoming increasingly difficult to produce enough food and to store it for long enough to supply everybody with a balanced diet. The science of **food technology** is using new developments to try and increase food supply.

Compare artificial selection with natural selection (see page 84).

The two processes have similarities but in natural selection it is change of the environment which is the selective agent.

Selective breeding

This is *selective* breeding to improve specific domesticated animals and crop plants. Important points are:

- people are the **selective agents** and choose the parent organisms which will breed
- the organisms are chosen because they have **desired characteristics**
- the aim is to incorporate the desired characteristics from both organisms in their offspring
- the offspring must be **assessed** to find out if they have the desired combination of improvements (there is **no guarantee** that a cross will be successful!)
- offspring which have suitable improvements are used for breeding, the others are deleted from the gene pool (not allowed to breed).

Most modern crops have been produced by artificial selection. The Brussels sprout variety below was produced in this way. Many trials were carried out before the new variety was offered for sale.

Can you suggest four excellent features offered by this new variety?

Brilliant NEW FOR 2001
F1 Hybrid A brand new early cropping variety which produces dense, dark green buttons of excellent quality in September and October. Suitable for a wide range of soil types it also has a high resistance to powdery mildew and ring spot. Good for freezing.   2152 pkt £2.10

Fertilisers

It is important that crop plants have access to all the minerals they require to give a maximum yield. Farmers supply these minerals in fertilisers, usually in the form NPK (nitrogen, phosphates and potassium). By supplying them with these minerals, nitrogen is available to make protein, a key substance for growth. Phosphates help the production of DNA, RNA and ATP. Potassium helps with protein synthesis and chlorophyll production. Other minerals are also needed like iron and calcium. The more a plant grows, the more its biomass increases and usually the greater the surface area for light absorption. The amount of photosynthesis increases proportionally. If a farmer is to reach the maximum productivity of a crop, fertiliser is vital.

Pesticides

If pests such as aphids or caterpillars begin to damage crops, then both quality and yield are reduced. Farmers combat pests by using chemical pesticides. Chemicals used to kill insects are insecticides. **Contact insecticides** kill insects directly but **systemic insecticides** are absorbed into the cell sap. Any insect consuming part of the plant or sucking the sap then dies.

Antibiotics

As well as treating crop plants with chemicals to improve yield, farmers can treat their animals with antibiotics. This is often done on a blanket basis not as a response to diseases. In this way, farmers hope to prevent disease in their animals and increase growth rates.

Microorganisms for food

Microorganisms are being increasingly used as a source of food. They include bacteria, fungi, yeasts and algae. These foods are often referred to as **single cell protein**. They have a number of advantages over more traditional sources of protein.

- They can be grown on waste products such as molasses and whey.
- They have a high growth rate.
- They do not need large areas of land for their cultivation.
- They often contain more fibre and less saturated fat than meat.
- They can be eaten by vegetarians.

There are disadvantages, however. They may need flavourings to make them more palatable and some are unsuitable for human consumption because of their high RNA content.

> Try to avoid talking about cost unless you qualify the statement. Single cell protein production plants are expensive to set up but they do run on inexpensive wastes.

Food preservation

In order to transport food long distances or store it for future consumption, methods of food preservation are used. This will prevent **microbial spoilage** by making the conditions unsuitable for microorganisms to grow:

- adding salt or sugar – this will reduce the water potential
- pickling – this reduces the pH
- freezing – this will reduce the temperature so that microbial enzymes are inactive
- heat treatment – pasteurisation to 70°C will kill harmful microorganisms and heating to higher temperatures will kill all microbes
- irradiation – exposure to gamma radiation will kill any microorganisms present and inhibit sprouting and ripening.

Coronary heart disease and lifestyle

OCR 2.2.1

Atherosclerosis is a major health problem caused by eating saturated fats. This circulatory disease may develop as follows:

- yellow fatty streaks develop under the lining of the **endothelium** on the inside of an artery
- the streaks develop into a fatty lump called an **atheroma**

- the atheroma is made from **cholesterol** (taken up in the diet as well as being made in the liver)
- dense **fibrous tissue** develops as the atheroma grows
- the endothelial lining can **split**, allowing blood to contact the fibrous atheroma
- the damage may lead to a blood clot and an artery can be blocked.

collagen fibres

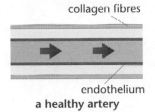 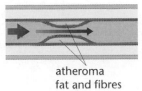 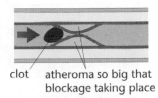

| endothelium | atheroma | clot atheroma so big that |
| **a healthy artery** | fat and fibres | blockage taking place |

Remember that the clotting of blood can occur for other reasons. There may be damage at other positions around the body. Blockage of this type is **thrombosis**.

Increasing constriction of an artery caused by **atherosclerosis** and **blood clots** reduces blood flow and increases blood pressure. If the artery wall is considerably weakened then a bulge in the side appears, just like a weakened inner tube on a cycle tyre. There is a danger of bursting and the structure is known as an **aneurysm**.

It is possible for a blood clot formed at an atheroma to break away from its original position. It may completely block a smaller vessel, this is known as an **embolism**.

Coronary heart disease

If the artery which supplies the heart (coronary artery) is partially blocked, then there is a reduction in oxygen and nutrient supply to the heart itself. This is called **coronary heart disease (CHD)**. The first sign is often **angina**, the main symptom being sharp chest pains. If total blockage occurs then **myocardial infarction** (heart attack) takes place.

Lifestyle and CHD

Many aspects of lifestyle influence the condition of the cardiovascular system. CHD is a multi-factorial disease and the risk of developing CHD depends on a number of factors.

Statistically people have a greater chance of avoiding CHD if they:

- consume a low amount of saturated fat in their diet
- do not drink alcohol excessively
- consume a low amount of salt
- do not smoke
- are not stressed most of the time
- exercise regularly.

New drugs called statins are being prescribed in an effort to reduce CHD. They are thought to work by decreasing LDL levels in the blood.

Saturated fats (see lipid structure pages 19–20) are found in large quantities in animal tissues. Eating large quantities of saturated fats seems to increase the risk of CHD. It seems that saturated fat increases the levels of **low-density lipoproteins** (LDLs) in the blood. LDLs transport cholesterol in the blood and high levels of LDLs seem to increase the risk of atheroma formation. Other fats such as polyunsaturated fats seem to increase the levels of **high-density lipoproteins** (HDLs). This seems to give some protection against CHD.

Exercise has a **protective effect** on the **heart** and **circulation**. Activities such as jogging, walking, swimming and cycling can:

- reduce the resting heart rate
- increase the strength of contraction of the heart muscle
- increase the stroke volume of the heart (the volume of blood which is propelled during the contractions of the ventricles).

High salt levels have been shown to increase blood pressure, therefore increasing the risk of damage to atheromas.

Smoking also increases blood pressure and makes the blood more likely to clot.

Progress check

(a) Name a specific substance in food which can result in atherosclerosis.

(b) Describe and explain the structural changes which take place in a blood vessel as atherosclerosis develops.

(c) How can the damage caused by an atheroma result in a heart attack?

(a) saturated fat

(b) yellow fatty streaks develop under the cells lining the inside of a blood vessel, the streaks develop into a lump known as an atheroma, the atheroma is made of cholesterol, dense fibrous tissue develops and the lining of the vessel can split.

(c) a blood clot forms which can block the blood vessel completely. Prevention of oxygen supply to the heart results in myocardial infarction (heart attack).

What are the dangers of smoking tobacco?

OCR 2.2.2

Each person has another choice to make, to smoke or not to smoke. The government health warning on every cigarette packet informs of health dangers but many young people go ahead and ignore the information.

Effects of tobacco smoking

- **Nicotine** is the active component in tobacco which **addicts** people to the habit.

- **Tars** coat the alveoli which **slows down exchange** of carbon dioxide and oxygen. If less oxygen is absorbed then the smokers will be less active than their true potential.

- **Cilia** lining bronchial tubes are coated then **destroyed**, reducing the efficiency in getting rid of pollutants which enter the lungs. These pollutants include the cigarette chemicals themselves.

- **Carbon monoxide** from the cigarette gases **combines with haemoglobin** of red blood cells rather than oxygen. This **reduces oxygen transport** and the smokers become less active than their potential. Ultimately it may lead to heart disease.

- The **bronchi** and **bronchioles** become **inflamed**, a condition known as **bronchitis**. This causes irritating fluid in the lungs, **coughing** and increased risk of heart disease. A number of bronchitis sufferers die each year.

- The walls of the alveoli break down reducing the surface area for gaseous exchange. Less oxygen can be absorbed by the lungs, leaving the **emphysema** sufferers extremely breathless. They increase their breathing rate to compensate but still cannot take in enough oxygen for a healthy life. Chronic emphysema sufferers need an oxygen cylinder to prolong their life.

- **Blood vessel elasticity is reduced** so that serious damage may occur. Ultimately a heart attack can follow.

- The **carcinogens** (cancer-causing chemicals) of the tobacco can result in **lung cancer**. **Malignant growths** in the lungs develop uncontrollably and cancers may spread to other parts of the body. Death often follows. Smokers have a greater risk of developing other cancers than non-smokers, e.g. more smokers develop cervical cancer.

Even non-smokers can develop any of the above symptoms, but the probability of developing them is increased by smoking. Being in a smoky atmosphere each day also increases the chances.

Learn the characteristics of each disease carefully. There are so many consequences of smoking that you may well mix them up.

The role of statistics

The government health warning on cigarette packets informs people of the risks of smoking. The WHO (World Health Organisation), governments and local authorities have collected statistics on many diseases over the years. These are used in education packs and posters to warn of risk factors. People can take precautions and use the information to avoid health dangers and take advantage of vaccination programmes.

Where education is not successful then related diseases follow. This acts as a drain on the National Health Service. Many operations which would have been unnecessary are performed to save people's lives, e.g. where coronary blood vessels are dangerously diseased a by-pass operation is the answer. The ideal situation is that education is successful, but realistically the aim is to balance prevention and cure.

8.2 Disease

After studying this section you should be able to:

LEARNING SUMMARY

- *define disease*
- *recall the causes, symptoms and control of a range of diseases including cholera, tuberculosis, malaria and AIDS*

What is a disease?

OCR 2.2

A disease is a **disorder** of a tissue, organ or system of an organism. As a result of a disorder, **symptoms** are evident. Such symptoms could be the failure to produce a particular digestive enzyme, or a growth of cells in the wrong place. Normal bodily processes may be disrupted, e.g. efficient oxygen transport is impeded by the malarial parasite, *Plasmodium*.

Different types of disease

Infectious disease by pathogens

Pathogens cause disease and can be passed from one person to another. Many pathogens are spread by a **vector** which carries them from one organism to another without being affected itself by the disease. Pathogens include bacteria, viruses, protozoa, fungi, parasites and worms.

Genetic diseases

These can be passed from parent to offspring, e.g. haemophilia and cystic fibrosis.

Dietary related diseases

These are caused by the foods that we eat. Too much or too little food may cause disorders, e.g. obesity or anorexia nervosa. Lack of vitamin D causes the bone disease rickets, the symptoms of which are soft weak bones which bend under the body weight.

Environmentally related diseases

Some aspects of the environment disrupt bodily processes, e.g. as a result of nuclear radiation leakage, cancer may develop.

Auto-immune disease

The body in some way attacks its own cells so that processes fail to function effectively.

> Most exam candidates recall that pathogens are responsible for disease. However, there are more causes of disease! If a question asks for different types of disease then giving a range of pathogens will not score many marks. Give genetic diseases, etc.

> An **auto-immune disease** may be **environmentally caused**, e.g. the form of leukaemia where phagocytes destroy a person's red blood cells may be caused by radiation leakage.

How are infectious diseases transmitted?

The pathogens which cause infectious diseases are transmitted in a range of ways.

- Direct contact – sexual intercourse enables the transmission of syphillis bacteria; a person's foot which touches a damp floor at the swimming baths can transfer the Athlete's foot fungus.

Be prepared to answer questions about diseases not on your syllabus. The examiners will give data and other information which you will need to interpret. Use your knowledge of the principles of disease transmission, infection, symptoms and cure.

- Droplet infection – a sneeze propels tiny droplets of nasal mucus carrying viruses such as those causing influenza.
- Via a vector – if a person with typhoid bacteria in the gut handles food the bacteria can be passed to a susceptible person.
- Via food or water – chicken meat kept in warm conditions encourages the reproduction of *Salmonella* bacteria which are transferred to the human consumer, who has food poisoning as a result.
- Via blood transfusion – as a result of receiving infected blood a person can contract AIDS.

Some infectious diseases have serious consequences to human life. The incidence of infectious diseases may vary according to the climate of the country, the presence of vectors, the social behaviour of people and other factors.

Disease file – tuberculosis

OCR   2.2.2

Cause of disease

Mycobacterium tuberculosis (bacterium) via droplet infection.

Transmission of microorganism

Coughs and sneezes of sufferers spread tiny droplets of moisture containing the pathogenic bacteria. People then inhale these droplets and may contract the disease.

Outline of the course of the disease and symptoms

The initial attack takes place in the lungs. The alveoli surfaces and capillaries are vulnerable and lesions occur. Some epithelial tissues begin to grow in number but these cannot carry out gaseous exchange. Inflammation occurs which stimulates painful coughing. Intense coughing takes place which can cause bleeding. There is much weight loss. Weak groups of people, like the elderly, or those underweight are more prone to the disease.

Prevention

Mycobacterium bovis causes tuberculosis in cattle. It can be passed to humans via milk. It causes an intestinal complaint in humans. It is important that cows are kept free of *M. bovis* by antibiotics.

The BCG vaccination is the injection of a weakened form of this microbe. This vaccination stimulates antibodies which are effective against both *M. tuberculosis* and *M. bovis*.

Mass screening using **X-rays** can identify 'shadows' in those people with scar tissue in the lungs.

Sputum testing identifies the presence of the bacteria in sufferers. Sufferers can be treated with antibiotics. Once cured they cannot pass on the pathogen so an epidemic may be prevented.

Skin testing is used. Antigens from dead *Mycobacteria* are injected just beneath the skin. If a person has been previously exposed to the organism then the skin swells which shows that they already have resistance, i.e. they have antibodies already. Anyone whose skin does not swell up is given the **BCG vaccination**. This contains attenuated *Mycobacterium bovis* to stimulate the production of antibodies against both *M. bovis* and *M. tuberculosis*.

Cure

Use of antibiotics such as streptomycin.

Disease file – malaria

OCR 2.2.2

Cause of disease

There are many variants of the malarial parasite, *Plasmodium* (protozoa). A parasite lives on in a living organism, causing it harm.

Transmission of the microorganism

The vector which carries the *Plasmodium* is a female *Anopheles* mosquito. The mosquito feeds on a mammal which may be suffering from malaria. It does this at night by inserting its 'syringe-like' stylet into a blood vessel beneath the skin. The mosquito feeds on blood and digests the red blood cells which releases the malarial parasites. These burrow into the insect's stomach wall where they breed; some then move to the salivary glands. Next time the mosquito feeds it secretes saliva to prevent clotting of the blood. This secretion introduces the parasites into the person's blood, who is likely to contract the disease.

Outline of the course of the disease and symptoms

After entry into the blood, **sporozoites** invade the **liver** releasing many **merozoites**. Each merozoite infects a **red blood cell** producing even more merozoites. Millions of these parasites are released into the blood causing a fever. As a result, the sufferer develops a range of symptoms including pains, exhaustion, aching, feeling cold, sweating and fever. The increased body temperature attracts mosquitoes even more, so a person with malaria acts as a reservoir for parasites.

Prevention

The most effective methods of prevention are those which **destroy the vector**. Spraying **insecticide** onto lake surfaces kills mosquito larvae.

Oil poured on lake surfaces prevents air entering the breathing tubes of the mosquitoes, so they die. Fish can be introduced into lakes as **predators** to eat the larvae. This is an example of **biological control**.

Sometimes ponds are **drained** to remove the mosquitoes' breeding area. People in areas where malaria is endemic **cover up** all waste **tin cans** and **plastic containers**. If they were to fill up with rain water then the mosquitoes have another habitat to breed in. The bacterium, ***Bacillus thuringiensis*** is used to destroy mosquitoes. Mosquito nets exclude mosquitoes from buildings and are even used over beds. **Electronic insect killer** techniques can be used which attract the mosquito via ultra-violet light then kill them by application of voltage. **Drugs** are used so that even if a person is bitten by a mosquito any *Plasmodia* entering the blood fail to develop further.

Combinations of these tests are used in different countries. Where there are outbreaks of the disease the systems are activated.

Cure

It is necessary to isolate and treat the sufferer. This also reduces the spread of the disease. Drugs are used to kill the parasites in the blood and reduce the symptoms. People are constantly attempting to find different ways of preventing this killer disease.

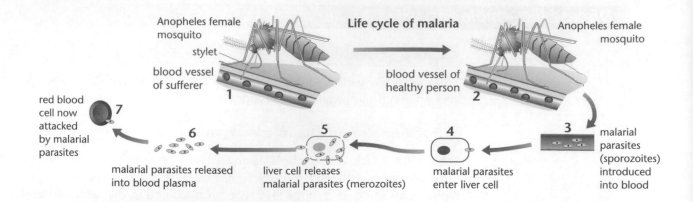

Life cycle of malaria

Anopheles female mosquito
stylet
blood vessel of sufferer
1

Anopheles female mosquito
blood vessel of healthy person
2

3 malarial parasites (sporozoites) introduced into blood

4 malarial parasites enter liver cell

5 liver cell releases malarial parasites (merozoites)

6 malarial parasites released into blood plasma

7 red blood cell now attacked by malarial parasites

Progress check

A mosquito carries the malarial parasite, *Plasmodium*. The female mosquito feeds on a mammal by inserting its 'syringe-like' stylet into a blood vessel beneath the skin. The mosquito feeds on blood and digests the red blood cells releasing the malarial parasites. These burrow into the insect's stomach wall and breed there, then some move to the salivary glands. Next time the mosquito feeds it secretes saliva. The saliva introduces the parasites into the person's blood.

(a) (i) Which species of mosquito transmits malaria?
(ii) Which organism causes the disease, malaria?
(iii) Does every mosquito bite transmit malaria?

Give a reason for your answer.

(b) Suggest how to reduce the spread of malaria.

(a) (i) *Anopheles* (ii) plasmodium (iii) no – mosquito must feed on sufferer first.

(b) Drain ponds where the mosquitoes breed; kill the mosquitoes with insecticide; pour oil on ponds to kill the larvae; introduce insectivorous fish as a form of biological control; use drugs such as chloroquine to cure people suffering from the disease; isolate people suffering from the disease; spray mosquitoes with a suspension of *Bacillis thuringiensis*; use of preventative anti-malarial drugs.

Disease file – AIDS (Acquired Immune Deficiency Syndrome)

OCR   2.2.2

Cause of disease

This is caused by HIV (human immune deficiency virus). It is a retrovirus, which is able to make DNA with the help of its own core of RNA.

Transmission of microorganism

This takes place by the exchange of body fluids, transfusion of contaminated blood, or via syringe needle 'sharing' in drug practices.

Outline of the course of the disease and symptoms

Destruction of T-lymphocyte cells

> Scientists are constantly trying to find a **cure**. None has been found yet.

The HIV protein coat attaches to protein in the plasma membrane of a T-lymphocyte. The virus protein coat fuses with the cell membrane releasing RNA and reverse transcriptase into the cell. This enzyme causes the cell to produce DNA from the viral RNA. This DNA enters the nucleus of the T-lymphocyte and is incorporated into the host cell chromosomes. The gene representing the HIV virus is permanently in the nucleus from now on and can be dormant for years. It may become activated by an infection. Viral protein and viral RNA are made as a result of the infection.

Many RNA viral cores now leave the cell and protein coats are assembled from degenerating plasma membranes. Other T-lymphocytes are attacked. Cells of the lymph nodes and spleen are also destroyed. Viruses appear in the blood, tears, saliva, semen and vaginal fluids. The immune system becomes so weak that many diseases can now successfully invade the weakened body.

Prevention

Screening of blood before transfusions. Use of condoms and remaining with one partner. No use of contaminated needles.

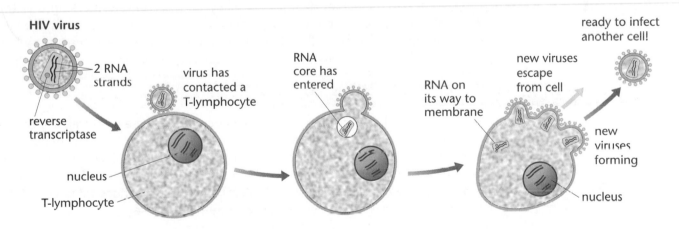

HIV virus — 2 RNA strands — reverse transcriptase — nucleus — T-lymphocyte — virus has contacted a T-lymphocyte — RNA core has entered — RNA on its way to membrane — new viruses escape from cell — new viruses forming — nucleus — ready to infect another cell!

8.3 Immunity

After studying this section you should be able to:

- describe the mechanisms used by the body to try and prevent pathogen entry
- describe and explain the action of the body's immune system

LEARNING SUMMARY

Survival against the attack of pathogens

OCR   2.2.2

Many pathogenic organisms attack people. They are not all successful in causing disease. We have immunity to a disease when we are able to resist infection. The body has a range of ways to prevent the disease-causing organism from becoming established.

- A tough protein called **keratin** helps skin cells to be a formidable **barrier** to prevent pathogens entering the body.
- An enzyme, **lysozyme**, destroys some microorganisms and can be found in sebum, tears and saliva.
- **Hydrochloric acid** in the stomach kills some microorganisms.
- The bronchial tubes of the lungs are lined with **cilia**. Microorganisms which enter the respiratory system are often trapped in mucus which is then moved to the oesophagus. From here they move to the stomach where many are destroyed by hydrochloric acid or digested.
- **Blood clotting** in response to external damage prevents entry of microorganisms from the external environment.

The methods the body uses to prevent microorganisms entering the bloodstream are sometimes unsuccessful. When the microorganisms invade and then breed in high numbers, we develop the **symptoms**. **White blood cells** enable us to destroy invading microorganisms. They may destroy the microorganisms quickly before they have any chance of becoming established, so the person would not develop any symptoms. Sometimes there are so many microorganisms attacking that the white blood cells cannot destroy all of them. Once the pathogens are established the symptoms of a disease follow, but for most diseases, after some time, the white blood cells eventually overcome the disease-causing organisms.

The roles of the white blood cells (leucocytes)

OCR   2.2.2

There are a number of different types of **leucocytes**. Most are produced from **stem cells** in the **bone marrow**. Different stem cells follow alternative maturation procedures to produce a range of leucocytes. Leucocytes have the ability to recognise self chemicals and non-self chemicals. Only where non-self chemicals are recognised will a leucocyte respond. Proteins and polysaccharides are typical of the complex molecules which can trigger an immune response.

Phagocytes

Phagocytes can move to a site of infection through capillaries, tissue fluid and lymph as well as being found in the plasma. They move towards pathogens which they destroy by the process of phagocytosis. This is often called engulfment and involves the surrounding of a pathogen by pseudopodia to form a food vacuole. Hydrolytic enzymes from lysosomes complete the destruction of the pathogen.

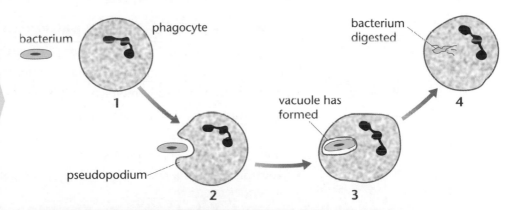

Neutrophils are one type of phagocyte. Proteins in plasma called **opsonins** attach to a pathogen. These opsonins enable the phagocyte to engulf the pathogen.

Macrophages are another type of phagocyte which work alongside T-lymphocytes.

What is an antigen?

OCR   2.2.2

As an individual grows and develops, complex substances such as proteins and polysaccharides are used to form cellular structures. Leucocytes identify these substances in the body as 'self' substances. They are ignored as the leucocytes encounter them daily. 'Non-self substances', e.g. foreign proteins which enter the body, are immediately identified as 'non-self'. These are known as **antigens** and trigger an immune response.

Lymphocytes

White blood cells (leucocytes) constantly check out proteins around the body. Foreign protein is identified and attack is stimulated.

In addition to phagocytes, there are other leucocytes called lymphocytes. There are two types of lymphocyte, **B-lymphocytes** and **T-lymphocytes**.

B-lymphocytes begin development and mature in the bone marrow. They produce antibodies, known as the **humoral response**.

T-lymphocytes work alongside phagocytes known as macrophages; this is known as the **cell-mediated response**. A macrophage engulfs an antigen. This antigen remains on the surface of the macrophage. T-lymphocytes respond to the antigen, dividing by mitosis to form a range of different types of T-lymphocyte cells.

- **Killer T-lymphocytes** adhere to the pathogen, secrete a toxin and destroy it.

- **Helper T-lymphocytes** stimulate the production of antibodies.

- **Suppressor T-lymphocytes** are inhibitors of the T-lymphocytes and plasma cells. Just weeks after the initial infection, they shut down the immune response when it is no longer needed.

- **Memory T-lymphocytes** respond to an antigen previously experienced. They are able to destroy the same pathogen before symptoms appear.

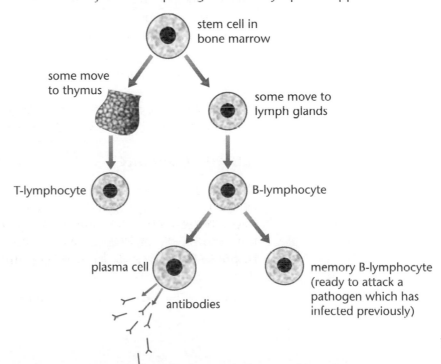

antigen

bacterium

flexible protein

antibody

antigen binding site

antibodies bind with antigens

How do antibodies destroy pathogens?

The diagram on the left shows antibodies binding to antigens. The descriptions below show what can happen immediately after the binding takes place.

There are three main ways in which antibodies destroy pathogens.

- **Precipitation**, by linking many antigens together. This enables the phagocytes to engulf them.
- **Lysis**, where the cell membrane breaks open, killing the cell.
- **Neutralisation** of a chemical released by the pathogen, so that the chemical is no longer toxic.

8.4 Assisting immunity

After studying this section you should be able to:

- *describe the origin of different types of immunity*
- *explain how immunity can be stimulated by vaccinations*

LEARNING SUMMARY

Stimulating the immune system

OCR 2.2.2

Newborn babies are naturally protected against many diseases such as measles and poliomyelitis. This is because they have received antibodies from their mother in two ways:

- across the placenta
- via breast feeding.

This type of immunity is called **passive natural immunity**. It is possible to give people injections of antibodies that have been made by another person or animal. This type of immunity is called **passive artificial** immunity.

> This is one major advantage of breast feeding over bottle feeding of babies.

> Passive immunity does not last very long. This is because the antibodies do not persist in the blood for very long.

KEY POINT

Active immunity

Once a person has recovered from certain diseases, e.g. measles, they rarely contract that disease again. This is because the memory cells formed from B-lymphocytes can survive in the blood for many years. If the antigen reappears, they can rapidly produce a clone of antibody producing cells. This **secondary response** is rapid and larger than the **primary response** and the antigen is rapidly destroyed.

This type of immunity is called **active natural immunity**.

Vaccines consist of dead or weakened (**attenuated**) forms of pathogens. They stimulate the production of memory cells so that the person develops active artificial **immunity**.

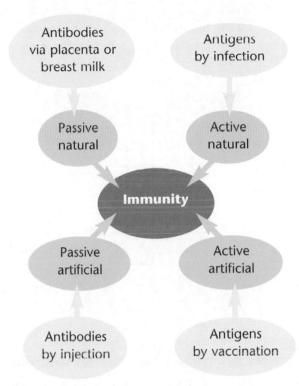

Unfortunately some pathogens show **antigenic variation**. This change in antigens renders vaccinations ineffective after a while and allows diseases such as influenza to strike many times in slightly different forms.

Scientists are now using antibody producing cells that have been fused with tumour cells. This produces cells that make large quantities of one type of antibody. They are called monoclonal antibodies. Because of their ability to target particular antigens, they might be useful in delivering drugs straight to target cells.

Sample question and model answers

The graph below shows the relative numbers of antibodies in a person's blood after the vaccination of attenuated viruses. Vaccinations were given on day 1 then 200 days later.

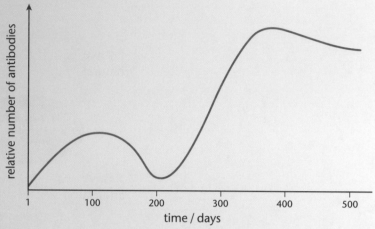

In examinations you are regularly given graphs. Make sure that you can link the idea being tested. This should help you recall all of the important concepts needed. All you need to do after this is **apply** your knowledge to the given data.

(a) Why is it important that viruses used in vaccinations are attenuated? [1]

If they were active then the person would contract the disease.

(b) Suggest **two** advantages of giving the second vaccination. [2]

A greater number of antibodies were produced.
The antibodies remain for much longer after the second vaccination.

(c) Which cells produced the antibodies during the primary response? [1]

B-lymphocytes.

(d) Why was there no delay in the secondary response to vaccination? [3]

Because the first vaccination had already been given, memory B-lymphocytes had been produced which respond to the viruses more quickly.

(e) Describe how a virus stimulates the production of antibodies? [3]

Antigen in the protein 'coat' or capsomere stimulate the B-lymphocytes.

If you gave B-lymphocytes as a response it would be wrong! B-lymphocytes secrete antibodies.

(f) Apart from producing antibodies, outline FOUR different ways that the body uses to destroy microorganisms. [4]

Phagocytes by engulfment; T-lymphocytes attach to microorganisms and destroy them; hydrochloric acid in the stomach; lysozyme in tears.

Practice examination questions

1 The illustration shows how the bacterium which causes TB can be transmitted from one person to another.

(a) Name the method of disease transmission shown in the illustration. [1]

(b) Sometimes the bacteria infect people but they do not develop symptoms.

   (i) What term is given to this group of people? [1]

   (ii) Explain why these people may be a greater danger to a community than those who actually suffer from the disease. [2]

(c) (i) What can be given to a person infected with TB to help destroy the bacteria? [1]

   (ii) Explain the role of each of the following in destroying TB bacteria.

   Phagocyte
   B-lymphocyte
   T-lymphocyte [6]

2 The diagram shows the response of B-lymphocytes to a specific antigen.

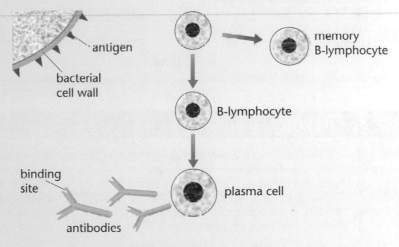

(a) (i) A plasma cell is bigger than a B-lymphocyte.
   Suggest an advantage of this. [1]

   (ii) Describe the precise role of antibodies in the immune response. [3]

   (iii) What is the advantage of memory B-lymphocytes? [2]

(b) What is an auto-immune disease? Give an example. [2]

(c) A person contracts the virus which causes the common cold.
   Suggest why their lymphocytes may fail to destroy the pathogen. [1]

Practice examination answers

Chapter 1 Biological molecules

1 (a) RCOOH   HOCH$_2$
        |
     RCOOH + HOCH
        |
     RCOOH   HOCH$_2$
     fatty acids glycerol   [2]

   (b) Emulsion test: add the sample to ethanol and mix;
       decant or pour into water;
       if a fat is present a white emulsion forms on the
       surface.   [3]
                                          [Total: 5]

2 (a) peptide bond/peptide link   [1]

   (b) –COOH/carboxylic acid   [1]

   (c) primary structure; amino acids in a chain   [2]
                                          [Total: 4]

3 (a) The latent heat of evaporation is large so lots of
       energy is needed to evaporate water/in sweating,
       much body heat is needed for evaporation.   [2]

   (b) High specific heat capacity means that the water
       needs a lot of heat energy to increase temperature
       significantly, therefore an organism will not
       overheat easily.   [2]

   (c) Cohesive forces aid the movement of water up the
       xylem.   [2]
                                          [Total: 6]

Chapter 2 Cells

1 (a) A = phospholipid B = protein   [2]

   (b) Acts as a channel/pore; to allow molecules in or out
       of the cell.   [1]

   (c) One end is hydrophobic and the other is
       hydrophilic; Hydrophilic tails 'hide' in the centre
       of the membrane.   [2]
                                          [Total: 5]

2 (a) C = Golgi body
       B = Centrioles
       A = Cell membrane
       E = Mitochondria
       D = Rough endoplasmic reticulum   [5]

   (b) Correct measure of width nucleus (~2.0 cm)
       convert to micrometres (20 000 µm)
       divide size by 5 000 (4 µm)   [3]

   (c) Liver cells carry out many functions;
       Need large amounts of energy/ATP   [2]
                                          [Total: 10]

3 (a) lens X = projector; lens Y = objective   [2]

   (b) electrons would collide with air molecules   [1]

   (c) (i) artefact   [1]
       (ii) ignore the artefact, because it is not part of
            normal structure and is only present
            due to preparation of the specimen.   [1]
                                          [Total: 5]

Chapter 3 Enzymes

1 (a) lock and key – the substrate is a similar shape to the
       active site; it fits in and binds with the active site
       like a key (substrate) fitting into a lock (active site);
       induced fit – the substrate is not a matching 'fit' for
       the active site, but as the substrate approaches, the
       active site changes into an appropriate shape.

   (b) reversible – enters active site but will come out again
       irreversible – binds permanently with enzymes [4]
                                          [Total: 4]

2 (a) starch   [1]

   (b) thermostable enzymes are effective at high
       temperatures   [1]
                                          [Total: 2]

3 (a) The substrate molecule collides with the active site
       of the enzyme; as it approaches, the active site
       changes shape to become compatible with the
       substrate shape.   [2]

   (b) Non-competitive inhibitor molecule binds with
       part of enzyme other than the active site; as a
       result the active site changes shape; so the
       substrate can no longer bind with the active
       site.   [3]
                                          [Total: 5]

4 (a) NH$_2$ (amino group)   [1]

   (b) The enzymes then have more ends to attack.   [2]
                                          [Total: 3]

Chapter 4 Exchange

1 (a) (i) No change in size because the water potential inside the cell equals the water potential of the solution outside the cell. [1]
   (ii) The water potential of the solution outside the cell is more negative than the water potential inside the cell. [1]
   (iii) The water potential of the solution outside the cell is less negative than the water potential inside the cell. [1]
   (b) osmosis [1]
   [Total: 4]

2 (a) They both use a protein carrier molecule. [1]
   (b) Active transport needs energy or mitochondria, whereas facilitated diffusion does not.
   OR Active transport allows molecules to move from a lower concentrated solution to a higher concentrated solution.
   OR Active transport allows molecules to move against a concentration gradient. [1]
   [Total: 2]

3 Alveoli have a very high surface area; they are very close to many capillaries; capillaries are one cell thick/very thin/have squamous epithelia; they are kept damp which facilitates diffusion. [4]
   [Total: 4]

Chapter 5 Transport

1 (a) to the body core [1]
   (b) less blood reaches the superficial capillaries of the skin; so less heat is lost by conduction, convection and radiation; blood in the body core better insulated by the adipose layer of the skin [4]
   [Total: 5]

2 (a) 60 x 120 x 5 = 36000 ml / 36 litres [2]
   (b) • blood is transported more quickly;
   • more oxygen taken up at the lungs/more carbon dioxide excreted at the lungs;
   • more oxygen reaches the muscles;
   • more glucose reaches the muscles;
   • so muscles contract more effectively.
   [any 4 points] [4]
   (c) (i) • slower breathing rate;
   • more alveoli accessed for exchange;
   • intercostal muscles more effective.
   [any 2 points] [2]
   (ii) • improved muscle tone;
   • greater muscular strength;
   • more capillaries in muscles. [any 2 points] [2]
   [Total: 10]

3 (a) The amount of water lost by transpiration is exactly matched by the amount taken up by the leaf. [1]
   (b) (i) xylem [1]
   (ii) fill the potometer under water; operate the valve to get rid of air bubbles; when removing the leaf from the tree, the stalk or petiole must be put in water immediately. [2]
   (c) Volume of water = πr^2 x 32 mm x 60
   = $\frac{22}{7}$ x 1 x 1 x 32 x 60
   = 6034.3 mm^3 [3]
   [Total: 7]

4 • extensive root system to absorb maximum water;
   • large amount of water storage in leaves;
   • thick cuticle;
   • low numbers of stomata/sunken stomata;
   • hairs to reduce turbulence. [any 3 points] [3]
   [Total: 3]

5 (a) SAN/Sinoatrial node [1]
   (b) (i) slows heart rate [1]
   (ii) speeds up the rate [1]
   (iii) speeds up the rate [1]
   [Total: 4]

6 (a) (i) sieve tube [1]
   (ii) companion cell has nucleus plus ribosomes; which make the proteins or enzymes; supplies enzymes to sieve tube via plasmodesmata. [3]
   (b) (i) It is an active process; a pump is involved; phloem contents under high pressure. [3]
   (ii) no starch in the phloem contents; did not change to blue-black; contained reducing sugar; did change to brick red. [4]
   (iii) Hot wax kills phloem cells; so they cannot transport the radioactive carbohydrate; transport by phloem is an active process. [3]
   [Total: 14]

Chapter 6 Genes and cell division

1 (a) metaphase   [1]
   (b) 4   [2]
                                                  [Total: 3]

2 (a) adenine and thymine are similar proportions
        because adenine binds with thymine;
        cytosine and guanine are similar proportions
        because cytosine binds with guanine   [2]

   (b) They should be identical in number but the
        scientists were operating at the limits of
        instrumentation.   [1]

   (c) Organic bases form the codes for different amino
        acids. Different sequences of amino acids form the
        different proteins specific to a species.   [2]
                                                  [Total: 5]

Chapter 7 Classification and biodiversity

1 A = Prokaryotae
   B = Fungi
   C = Protoctista
   D = Plantae
   E = Animalia   [5]
                                                  [Total: 5]

2 (a)

	Mistle-thrush	Song-thrush
Kingdom	**Animalia**	**Animalia**
Phylum	Chordata	Chordata
Class	Aves	Aves
Order	Passeriformes	Passeriformes
Family	**Turdidae**	**Turdidae**
Genus	**Turdus**	**Turdus**
Species	**viscivorus**	**ericetorum**

                                                       [3]
   (b) disruptive selection   [1]
                                                  [Total: 4]

3 (a) N = 20

$$D = 1 - [(\frac{3}{20})^2 + (\frac{6}{20})^2 + (\frac{4}{20})^2 + (\frac{2}{20})^2 + (\frac{1}{20})^2 + (\frac{1}{20})^2 + (\frac{3}{20})^2]$$

$$= 1 - [0.023 + 0.09 + 0.04 + 0.01 + 0.0025 + 0.0025 + 0.023]$$

$$= 1 - 0.19$$

$$= 0.81 \qquad [2]$$

   (b) Biodiversity was lower in pond near feeding station.
        Difference may not be significant.   [2]
                                                  [Total: 4]

Chapter 8 Human health and disease

1 (a) droplet infection   [1]
   (b) (i) carriers   [1]
        (ii) we do not know that they carry the pathogen
             as they display no symptoms, so the people do
             not avoid contact and pass on the bacteria   [2]
   (c) (i) antibiotics or named antibiotics   [1]
        (ii) **Phagocyte** – engulfs/produces pseudopodia/
             phagocytosis; digests the bacterium/causes lysis
             of the bacterium   [2]
             B-lymphocyte – changes into plasma cell;
             makes antibodies   [2]
             T-lymphocyte – whole cell links to bacterial
             antigen sites; cell is usually destroyed
             by this; reacts to bacterial antigen   [2]
                                                  [Total: 11]

2 (a) (i)   ability to secrete more antibodies   [1]
        (ii) the antibodies have specific receptor sites
              which bind with the antigens; they have
              a flexible protein which changes angle to fit
              the antigens; antibodies result in the
              destruction of the antigen in some way/
              neutralise toxin/cluster around antigens then
              cause precipitation/cause agglutination   [3]
        (iii) are produced when body first exposed to
              antigen; remain in body to react quickly
              when exposed to same antigen again   [2]
   (b) when the immune system attacks the person's own
        cells; pernicious anaemia/rheumatoid arthritis   [2]
   (c) the influenza virus often mutates so lymphocytes
        take longer to produce antibodies   [1]
                                                  [Total: 9]

Notes

Notes

Index

Index

Revise
A2

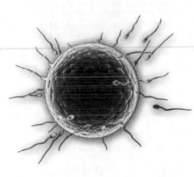

OCR
Biology

John Parker & Ian Honeysett

Contents

Contents

Specification list

The Specification labels on each page refer directly to the units in the exam board specification, i.e. OCR 4.3.1 refers to unit 4, module 3, section 1.

OCR Biology

UNIT	SPECIFICATION TOPIC	CHAPTER REFERENCE	STUDIED IN CLASS	REVISED	PRACTICE QUESTIONS
Unit 4 (M1)	Communication	3.1, 3.2			
	Nerves	2.1, 2.2, 2.3			
	Hormones	3.1, 3.3			
Unit 4 (M2)	Excretion	3.3, 3.4			
Unit 4 (M3)	Photosynthesis	1.2			
Unit 4 (M4)	Respiration	1.1, 1.3			
Unit 5 (M1)	Cellular control	4.1, 4.3			
	Meiosis and variation	4.2, 5.1, 5.2			
Unit 5 (M2)	Cloning in plants and animals	6.2			
	Biotechnology	6.1			
	Genomes and gene technology	6.3			
Unit 5 (M3)	Ecosystems	8.1, 8.2, 7.1, 7.2			
	Populations and sustainability	7.1			
Unit 5 (M4)	Plant responses	2.6			
	Animal responses	2.4, 2.5			
	Animal behaviour	2.5, 7.3			

Examination analysis

Unit 4
1 hour written examination   A Level – 15%

Unit 5
1 hour 45 mins written examination   A Level – 25%

Unit 6
Internal assessment   A Level – 10%

The AS/A2 Level Biology course

All Biology GCE A level courses currently studied are in two parts: AS and A2, with three separate units in each.

Some of the units are assessed by written papers, externally marked by the Awarding Body. Some units involve internal assessment of practical skills (subject to moderation).

Each Awarding Body has a common core of subject content in AS and A2. Beyond the common core material, the Awarding Bodies have included more varied content. This study guide contains the common core material and the additional material that is relevant to the OCR A2 specification.

In using this study guide, some students may have already completed the AS part of the course. Knowledge of AS is assumed in the A2 part of the course. It is therefore important to revisit the AS information when preparing for the A2 examinations.

What are the differences between AS and A2?

There are three main differences:

(i) A2 includes the more **demanding** concepts. (Understanding will be easier if you have completed the AS Biology course as a 'stepping stone'.)

(ii) There is a much greater emphasis on the skills of **application** and **analysis** than in AS. (Using knowledge and understanding acquired from AS is essential.)

(iii) A2 includes a substantial amount of **synoptic** material. (This is the drawing together of knowledge and skills across the modules of AS and A2. Synoptic investigative tasks and questions involving concepts across the specification are included.)

How will you be tested?

Assessment units

OCR A2 Biology comprises three units. The first two units are assessed by examinations.

The third component involves centre assessed practical assessment. This tests practical skills and is marked by your teacher. The marks can be adjusted by moderators appointed by OCR.

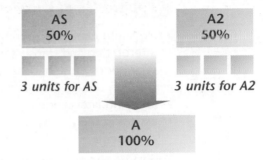

Tests are taken at two specific times of the year, January/February and June. If you are disappointed with a unit result, you can resit each unit any number of times. It can be an advantage to you to take a unit test at the earlier optional time because you can re-sit the test. The best mark from each unit will be credited and the lower marks ignored.

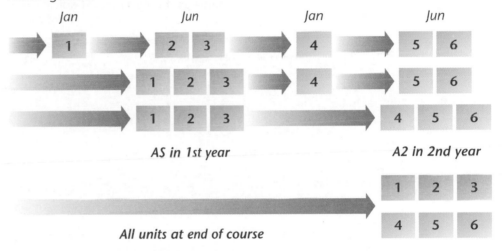

A2 and synoptic assessment

Most students who study A2 have already studied to AS Level. There are three further units to be studied.

Every A Level specification includes synoptic assessment at the end of A2. Synoptic questions draw on the ideas and concepts of earlier units, bringing them together in holistic contexts. Examiners will test your ability to inter-relate topics through the complete course from AS to A2. (See the synoptic chapter page 113).

What skills will I need?

For A2 Biology, you will be tested by assessment objectives: these are the skills and abilities that you should have acquired by studying the course. The assessment objectives are shown below.

Knowledge with understanding

- recall of facts, terminology and relationships
- understanding of principles and concepts
- drawing on existing knowledge to show understanding of the responsible use of biological applications in society
- selecting, organising and presenting information clearly and logically

Application of knowledge and understanding, and evaluation

- explaining and interpreting principles and concepts
- interpreting and translating, from one to another, data presented as continuous prose or in tables, diagrams and graphs
- carrying out relevant calculations
- applying knowledge and understanding to familiar and unfamiliar situations
- assessing the validity of biological information, experiments, inferences and statements

You must also present arguments and ideas clearly and logically, using specialist vocabulary where appropriate. Remember to balance your argument!

Experimental and investigative skills

One of the A2 units tests experimental and investigative skills. The format of this unit is the same as the AS practical unit.

Different types of questions in A2 examinations

Questions in AS and A2 Biology are designed to assess a number of assessment objectives. For the written papers in Biology the main objectives being assessed are:

- recall of facts, terminology and inter-relationships
- understanding of principles and concepts and their social and technological applications and implications
- explanation and interpretation of principles and concepts
- interpreting information given as diagrams, photomicrographs, electron micrographs tables, data, graphs and passages
- application of knowledge and understanding to familiar and unfamiliar situations.

In order to assess these abilities and skills a number of different types of question are used.

In A2 Level Biology unit tests these include short-answer questions and structured questions requiring both short answers and more extended answers, together with free-response and open-ended questions.

Short-answer questions

A short-answer question will normally begin with a brief amount of stimulus material. This may be in the form of a diagram, data or graph. A short-answer question may begin by testing recall. Usually this is followed up by questions which test understanding. Often you will be required to analyse data. Short-answer questions normally have a space for your responses on the printed paper. The number of lines is a guide as to the number of words you will need to answer the question. The number of marks indicated on the right side of the papers shows the number of marks you can score for each question part. Here are some examples. (The answers are shown in blue).

The diagram below shows a gastric pit.

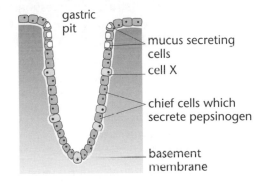

(a) (i) Label cell X   (1)
         oxyntic cell

   (ii) What is secreted by cell X?   (1)
         hydrochloric acid

(b) (i) Protein enters the stomach. What must take place before the hydrolysis of the protein begins?   (2)
         Hydrochloric acid acts on pepsinogen, to produce pepsin

   (ii) After the protein has been hydrolysed, what is produced?   (1)
         polypeptides

Structured questions

Structured questions are in several parts. The parts are usually about a common context and they often progress in difficulty as you work through each of the parts. They may start with simple recall, then test understanding of a familiar or unfamiliar situation. If the context seems unfamiliar the material will still be centred around concepts and skills from the Biology specification. (If a student can answer questions about unfamiliar situations then they display understanding rather than simple recall.)

The most difficult part of a structured question is usually at the end. Ascending in difficulty, a question allows a candidate to build in confidence. Right at the end technological and social applications of biological principles give a more demanding challenge. Most of the questions in this book are structured questions. This is the main type of question used in the assessment of both AS and A2 Biology.

The questions set at A2 Level are generally more difficult than those experienced at AS Level. A2 includes a number of higher-level concepts, so can be expected to be more difficult. The key advice given by this author is:

- Give your answers in greater detail:

  Example: Why does blood glucose rise after a period without food?

  Answer: The hormone glucagon is produced ✗ *(This is not enough for credit!)*

  The hormone glucagon is produced which results in glycogen breakdown to glucose. ✓

- Look out for questions with a 'sting in the tail'. A2 structured questions are less straightforward, so look for a 'twist'. This is identified in the example below.

When answering structured questions, do not feel that you have to complete one question before starting the next. Answering a part that you are sure of will build your confidence. If you run out of ideas go on to the next question. This will be more profitable than staying with a very difficult question which slows down progress. You can return at the end when you have more time.

Extended answers

In A2 and AS Biology, questions requiring more extended answers will usually form part of structured questions. They will normally appear at the end of a structured question and will typically have a value of 4 to 10 marks. Longer questions are allocated more lines, so you can use this as a guide as to how many points you need to make in your response. Often for an answer worth 10 marks the mark scheme would have around 12 to 14 creditable answers. You are awarded up to the maximum, 10 marks, in this instance.

Longer extended answers are used to allocate marks for the **quality of communication**.

Candidates are assessed on their ability to use a suitable style of writing, and organise relevant material, both logically and clearly. The use of specialist biological terms in context is also assessed. Spelling, punctuation and grammar are also taken into consideration. Here is a longer-response question.

Question

Urea, glucose and water molecules enter the kidney via the renal artery. Explain what *can* happen to each of these substances.

In this question one mark is available for communication. (Total 10 marks)

Urea, glucose and water molecules can pass out of the blood capillaries in a glomerulus. ✓ This is as a result of ultrafiltration, ✓ as the narrow diameter of the efferent blood vessel cause a pressure build up. ✓

The three substances pass down the proximal tubule. 100% glucose is reabsorbed in the proximal tubule ✓ so is returned to the blood. Carrier proteins on the microvilli aided by mitochondria, actively transport the glucose across the cells. ✓ Around 80% of the water is reabsorbed in the proximal tubule. ✓ Remaining water and urea molecules continue through the loop of Henlé. Urea continues through the distal tubule to the ureter then the bladder. ✓

More water can be reabsorbed with the help of the countercurrent multiplier. ✓ The ascending limb of the loop of Henlé ✓ actively transports Na^+ and Cl^- ions into the medulla. ✓ Water molecules leave the collecting duct by osmosis due to the ions in the medulla. ✓ Cells of the collecting duct are made more permeable to water by the hormone, ADH. ✓ Some water molecules pass into the capillary network and having been successfully reabsorbed. ✓
Some water molecules continue down the ureters and into the bladder. ✓

Communication mark ✓

Remember that mark schemes for extended questions often exceed the question total, but you can only be awarded credit up to the maximum. In response to this question the candidate would be awarded the maximum of 10 marks which included one communication mark. The candidate gave five more creditable responses which were on the mark scheme, but had already scored a maximum. Try to give more detail in your answers to longer questions. This is the key to A2 success.

Stretch and Challenge

Stretch and Challenge is a concept that is applied to the structured questions in Unit 4 and 5 of the exam papers in A2. In principle, it means that sub-questions become progressively harder so as to challenge more able students and help differentiate between A and A* students.

Stretch and Challenge questions are designed to test a variety of different skills and your understanding of the material. They are likely to test your ability to make appropriate connections between different areas and apply your knowledge in unfamiliar contexts (as opposed to basic recall).

Exam technique

A2 builds from the skills and concepts acquired during the AS course. It will help you cope as the A2 concepts ascend in difficulty. The chapters explain the ideas in small steps so that understanding takes place gradually. The final aim, of complete understanding of major topics, is more likely.

Can I use my AS Biology Study Guide for A2?

YES! This will be particularly useful in answering synoptic questions that require direct knowledge of the AS topics.

What are examiners looking for?

Whatever type of question you are answering, it is important to respond in a suitable way. Examiners use instructions to help you to decide the length and depth of your answer. The most common words used are given below, together with a brief description of what each word is asking for.

Define

This requires a formal statement. Some definitions are easy to recall.

Define the term transport.

This is the movement of molecules from where they are in lower concentration to where they are in higher concentration. The process requires energy.

Other definitions are more complex. Where you have problems it is helpful to give an example.

Define the term endemic.

This means that a disease is found regularly in a group of people, district or country.

Use of an example clarifies the meaning. Indicating that malaria is invariably found everywhere in a country confirms understanding.

Explain

This requires a reason. The amount of detail needed is shown by the number of marks allocated.

Explain the difference between resolution and magnification.

Resolution is the ability to be able to distinguish between two points whereas magnification is the number of times an image is bigger than an object itself.

State

This requires a brief answer without any reason.

State one role of blood plasma in a mammal.

Transport of hormones to their target organs.

List

This requires a sequence of points with no explanation.

List the abiotic factors which can affect the rate of photosynthesis in pondweed.

carbon dioxide concentration; amount of light; temperature; pH of water

Describe

This requires a piece of prose which gives key points. Diagrams should be used where possible.

Describe the nervous control of heart rate.

The medulla oblongata ✓ of the brain connects to the sino-atrial node in the right atrium, wall ✓ via the vagus nerve and the sympathetic nerve ✓ the sympathetic nerve speeds up the rate ✓ the vagus nerve slows it down. ✓

Discuss

This requires points both for and against, together with a criticism of each point. (**Compare** is a similar command word).

Discuss the advantages and disadvantages of using systemic insecticides in agriculture.

Advantages are that the insecticides kill the pests which reduce yield ✓ they enter the sap of the plants so insects which consume sap die ✓ the insecticide lasts longer than a contact insecticide, 2 weeks is not uncommon ✓

Disadvantages are that insecticide may remain in the product and harm a consumer e.g. humans ✓ it may destroy organisms other than the target ✓ no insecticide is 100% effective and develops resistant pests. ✓

Suggest

This means that there is no single correct answer. Often you are given an unfamiliar situation to analyse. The examiners hope for logical deductions from the data given and that, usually, you apply your knowledge of biological concepts and principles.

The graph shows that the population of lynx decreased in 1980. Suggest reasons for this.

Weather conditions prevented plant growth ✓ so the snowshoe hares could not get enough food and their population remained low ✓ so the lynx did not have enough hares (prey) to predate upon. ✓ The lynx could have had a disease which reduced numbers. ✓

Calculate

This requires that you work out a numerical answer. Remember to give the units and to show your working, marks are usually available for a partially correct answer. If you work everything out in stages write down the sequence. Otherwise if you merely give the answer and it is wrong, then the working marks are not available to you.

Calculate the Rf value of spot X. (X is 25 mm from start and solvent front is 100 mm)

$$Rf = \frac{\text{distance moved by spot}}{\text{distance moved by the solvent front}}$$

$$= \frac{25 \text{ mm}}{100 \text{ mm}} = 0.25$$

Outline

This requires that you give only the main points. The marks allocated will guide you on the number of points which you need to make.

Outline the use of restriction endonuclease in genetic engineering.

The enzyme is used to cut the DNA of the donor cell. ✓

It cuts the DNA up like this A T G C C G A T = A T   + G C C G A T ✓
                                T A C G G C T A   T A C G G C   T A

The DNA in a bacterial plasmid is cut with the same restriction endonuclease. ✓

The donor DNA will fit onto the sticky ends of the broken plasmid. ✓

If a question does not seem to make sense, you may have misread it. Read it again!

Some dos and don'ts

Dos

Do *answer the question*

No credit can be given for good Biology that is irrelevant to the question.

Do *use the mark allocation to guide how much you write*

Two marks are awarded for two valid points – writing more will rarely gain more credit and could mean wasted time or even contradicting earlier valid points.

Do *use diagrams, equations and tables in your responses*

Even in 'essay-style' questions, these offer an excellent way of communicating Biology.

Do *write legibly*

An examiner cannot give marks if the answer cannot be read.

Do *write using correct spelling and grammar. Structure longer essays carefully*

Marks are now awarded for the quality of your language in exams.

Don'ts

Don't *fill up any blank space on a paper*

In structured questions, the number of dotted lines should guide the length of your answer.

If you write too much, you waste time and may not finish the exam paper. You also risk contradicting yourself.

Don't *write out the question again*

This wastes time. The marks are for the answer!

Don't *contradict yourself*

The examiner cannot be expected to choose which answer is intended. You could lose a hard-earned mark.

Don't *spend too much time on a part that you find difficult*

You may not have enough time to complete the exam. You can always return to a difficult calculation if you have time at the end of the exam.

What grade do you want?

Everyone would like to improve their grades but you will only manage this with a lot of hard work and determination. You should have a fair idea of your natural ability and likely grade in Biology and the hints below offer advice on improving that grade.

For a Grade A

You will need to be a very good all-rounder.

- You must go into every exam knowing the work extremely well.
- You must be able to apply your knowledge to new, unfamiliar situations.
- You need to have practised many, many exam questions so that you are ready for the type of question that will appear.

The exams test all areas of the syllabus and any weaknesses in your Biology will be found out. There must be no holes in your knowledge and understanding. For a Grade A, you must be competent in all areas.

For a Grade C

You must have a reasonable grasp of Biology but you may have weaknesses in several areas and you will be unsure of some of the reasons for the Biology.

- Many Grade C candidates are just as good at answering questions as the Grade A students but holes and weaknesses often show up in just some topics.
- To improve, you will need to master your weaknesses and you must prepare thoroughly for the exam. You must become a better all-rounder.

For a Grade E

You cannot afford to miss the easy marks. Even if you find Biology difficult to understand and would be happy with a Grade F, there are plenty of questions in which you can gain marks.

- You must memorise all definitions.
- You must practise exam questions to give yourself confidence that you do know some Biology. In exams, answer the parts of questions that you know first. You must not waste time on the difficult parts. You can always go back to these later.
- The areas of Biology that you find most difficult are going to be hard to score on in exams. Even in the difficult questions, there are still marks to be gained. Show your working in calculations because credit is given for a sound method. You can always gain some marks if you get part of the way towards the solution.

What marks do you need?

The table below shows how your average mark is transferred into a grade.

average	80%	70%	60%	50%	40%
grade	A	B	C	D	E

The A* grade

To achieve an A* grade, you need to achieve a...

- grade A overall (80% or more on uniform mark scale) for the whole A level qualification
- grade A* (90% or more on the uniform mark scale) across your A2 units.

A* grades are awarded for the A level qualification only and not for the AS qualification or individual units.

Four steps to successful revision

Step 1: Understand

- Study the topic to be learned slowly. Make sure you understand the logic or important concepts.
- Mark up the text if necessary – underline, highlight and make notes.
- Re-read each paragraph slowly.

GO TO STEP 2

Step 2: Summarise

- Now make your own revision note summary:
  What is the main idea, theme or concept to be learned?
  What are the main points? How does the logic develop?
  Ask questions: Why? How? What next?
- Use bullet points, mind maps, patterned notes.
- Link ideas with mnemonics, mind maps, crazy stories.
- Note the title and date of the revision notes
  (e.g. Biology: Homeostasis, 3rd March).
- Organise your notes carefully and keep them in a file.

This is now in **short-term memory**. You will forget 80% of it if you do not go to Step 3.
GO TO STEP 3, but first take a 10 minute break.

Step 3: Memorise

- Take 25 minute learning 'bites' with 5 minute breaks.
- After each 5 minute break test yourself:
  Cover the original revision note summary.
  Write down the main points.
  Speak out loud (record on tape).
  Tell someone else.
  Repeat many times.

The material is well on its way to **long-term memory**.
You will forget 40% if you do not do step 4. *GO TO STEP 4*

Step 4: Track/Review

- Create a Revision Diary (one A4 page per day).
- Make a revision plan for the topic, e.g. 1 day later, 1 week later, 1 month later.
- Record your revision in your Revision Diary, e.g.
  Biology: Homeostasis, 3rd March 25 minutes
  Biology: Homeostasis, 5th March 15 minutes
  Biology: Homeostasis, 3rd April 15 minutes
  ... and then at monthly intervals.

Chapter 1

Energy for life

The following topics are covered in this chapter:

- Metabolism and ATP
- Respiration

- Autotrophic nutrition

1.1 Metabolism and ATP

After studying this section you should be able to:

- understand the principles of metabolic pathways
- understand the importance of ATP

Metabolic pathways

OCR 4.3.1, 4.4.1

Inside a living organism there are many chemical reactions occurring at the same time. They may be occurring in the same place, in different parts of the cell or in different cells. Each reaction is controlled by a different enzyme.

> **KEY POINT**
>
> All the chemical reactions occurring in an organism are called **metabolism**.

Often a number of chemical reactions are linked together. The product of one reaction acts as the substrate for the next reaction. This is called a **metabolic pathway** and each of the chemicals in the pathway are called **intermediates**.

A is the substrate for this pathway, B, C and D are intermediates and E is the product. The enzymes a, b, c and d each control a different step.

$$A \xrightarrow{a} B \xrightarrow{b} C \xrightarrow{c} D \xrightarrow{d} E$$

Metabolic reactions can be classed as one of two types. Reactions that break down complex molecules are called **catabolic reactions** or catabolism. Other reactions build up complex molecules from simple molecules. They are **anabolic reactions** or anabolism.

Anabolic reactions tend to require energy, whereas catabolic reactions release energy. The link between these two types of reactions is a molecule called **ATP**.

Adenosine triphosphate (ATP)

The breakdown of many organic molecules can release large amounts of energy. Similarly, making complex molecules such as proteins requires energy. These reactions must be coupled together. This is achieved by using **adenosine triphosphate (ATP)** molecules.

ATP is a **phosphorylated nucleotide**. (Recall the structure of DNA which consists of nucleotides.) Each nucleotide consists of an organic base, ribose sugar and phosphate group. ATP is a nucleotide with two extra phosphate groups. This is the reason for the term 'phosphorylated nucleotide'.

adenine ——— ribose ——— phosphate ——— phosphate ——— phosphate

ATP is produced from adenosine diphosphate and a phosphate group. This requires energy. The energy is trapped in the ATP molecule. An enzyme called ATPase catalyses this reaction.

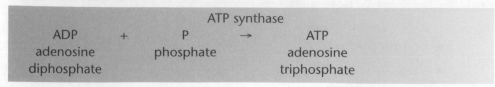

ATP synthase

| ADP | + | P | → | ATP |
| adenosine diphosphate | | phosphate | | adenosine triphosphate |

The hydrolysis of the terminal phosphate group liberates the energy. This can then be used in a number of different ways.

> ATPase is a hydrolysing enzyme so that a water molecule is needed, but this is not normally shown in the equation.

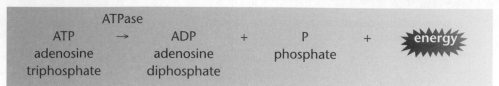

ATPase

| ATP | → | ADP | + | P | + | energy |
| adenosine triphosphate | | adenosine diphosphate | | phosphate | | |

> **KEY POINT**
>
> ATP is the cell's energy currency. A cell does not store large amounts of ATP but it uses it to transfer small packets of energy from one set of reactions to another.

Uses of ATP

- muscle contraction
- active transport
- synthesis of macromolecules
- stimulating the breakdown of substrates to make even more ATP for other uses

1.2 Autotrophic nutrition

After studying this section you should be able to:

- *describe the part played by chloroplasts in photosynthesis*
- *recall and explain the biochemical processes of photosynthesis*
- *relate the properties of chlorophyll to the absorption and action spectra*
- *understand how the law of limiting factors is linked to productivity*
- *understand that glucose can be converted into a number of useful chemicals*

LEARNING SUMMARY

Synthesising food

OCR 4.3.1

Autotrophic nutrition is very important. Autotrophic nutrition means that simple inorganic substances are taken in and used to synthesise organic molecules. Energy is needed to achieve this. In **photo-autotrophic nutrition** light is the energy source. In most instances, the light source is **solar energy**, the process being **photosynthesis**. Carbon dioxide and water are taken in by organisms and used to synthesise glucose, which can be broken down later during respiration to release the energy needed for life. Glucose molecules can be polymerized into starch, a storage substance. By far the greatest energy supply to support food chains and webs is obtained from photo-autotrophic nutrition. Most producers use this nutritional method.

The chloroplast

OCR 4.3.1

Chloroplasts are organelles in plant cells which photosynthesise. In a leaf, they are strategically positioned to absorb the maximum amount of light energy. Most are located in the palisade mesophyll of leaves, but they are also found in both spongy mesophyll and guard cells. There is a greater amount of light entering the upper surface of a leaf so the palisade tissues benefit from a greater chloroplast density.

The diagram below shows the structure of a chloroplast.

> Remember that not all light reaching a leaf may hit a chloroplast. Photons of light can be reflected or even absorbed by other parts of the cell. Around 4% of light entering an ecosystem is actually utilised in photosynthesis!
>
> Even when light reaches the green leaf, not all energy is fixed in the carbohydrate product. Just one quarter becomes chemical energy in carbohydrate.

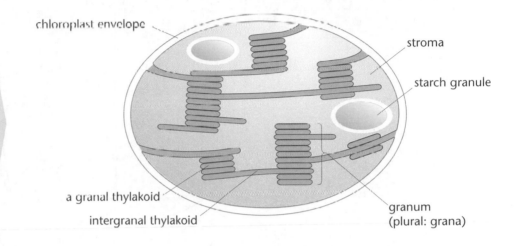

Structure and function

A system of **thylakoid membranes** is located throughout the chloroplast. These are flattened membranous vesicles which are surrounded by a liquid-based matrix, the **stroma**.

Along the thylakoid membranes are key substances:

* chlorophyll molecules
* other pigments
* enzymes
* electron acceptor proteins.

Throughout the chloroplasts, circular thylakoid membranes stack on top of each other to form **grana**. Grana are linked by longer **intergranal thylakoids**. Granal thylakoids and intergranal thylakoids have different pigments and proteins. Each type has a different role in photosynthesis.

The key substances in the thylakoids occur in specific groups comprising pigment, enzyme and electron acceptor proteins. There are two specific groups known as **photosystem I** and **photosystem II**.

> Do not be confused by the photosystems. They are groups of chemicals which harness light and pass on energy. Remember this information to understand the biochemistry of photosynthesis.

The photosystems

Each photosystem contains a large number of chlorophyll molecules. As light energy is received at the chlorophyll, electrons from the chlorophyll are boosted to a higher level and energy is passed to pigment molecules known as the **reaction centre**.

> **KEY POINT**
>
> The reaction centre of photosystem I absorbs energy of wavelength 700 nanometres. The reaction centre of photosystem II absorbs energy of wavelength 680–690 nanometres. In this way, light of different wavelengths can be absorbed.

The process of photosynthesis

OCR   4.3.1

The process of photosynthesis is summarised by the flow diagram below.

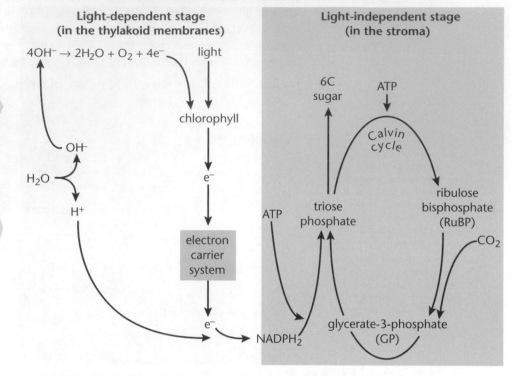

In examinations, look out for parts of this diagram. There may be a few empty boxes where a key substance is missing. Will you be able to recall it?

- Photosynthesis harnesses solar energy.
- Photosynthesis involves light-dependent and light-independent reactions.
- Photosynthesis results in the flow of energy through an ecosystem.

Light-dependent reaction

- Light energy results in the excitation of electrons in the **chlorophyll**.
- These electrons are passed along a series of electron acceptors in the thylakoid membranes, collectively known as the **electron carrier system**.
- Energy from excited electrons funds the production of **ATP** (adenosine triphosphate).
- The final electron acceptor is **NADP⁺**.
- Electron loss from chlorophyll causes the splitting of water (photolysis):

$$H_2O \rightarrow H^+ + OH^- \quad \text{then} \quad 4OH^- \rightarrow 2H_2O + O_2 + 4e^-$$

- Oxygen is produced, water to re-use, and electrons stream back to replace those lost in the chlorophyll.
- Hydrogen ions (H^+) from photolysis, together with $NADP^+$ form **NADPH$_2$**.

No ATP and NADPH$_2$ in a chloroplast would result in no glucose being made. Once supplies of ATP and NADPH$_2$ are exhausted then photosynthesis is ended. In examinations look out for the 'lights out' questions where the light-independent reaction continues for a while until stores of ATP, NADPH$_2$ and GP are used up. These questions are likely to be graph based.

Light-independent reaction

- Two useful substances are produced by the light-dependent stage, ATP and **NADPH$_2$**. These are needed to drive the light-independent stage.
- They react with glycerate-3-phosphate (GP) to produce a triose sugar – **triose phosphate**.
- Triose phosphate is used *either* to produce a 6C sugar *or* to form **ribulose bisphosphate** (RuBP).
- The conversion of triose phosphate (3C) to RuBP occurs in the Calvin cycle and utilises ATP, which supplies the energy required.
- A RuBP molecule (5C) together with a carbon dioxide molecule (1C) forms two GP molecules (2 × 3C) to complete the Calvin cycle.
- The GP is then available to react with ATP and NADPH$_2$ to synthesise more triose sugar or RuBP.

How do the photosystems contribute to photosynthesis?

This can be explained in terms of the **Z scheme** shown below.

> The **Z scheme**, so called because the paths of electrons shown in the diagram are in a 'Z' shape.

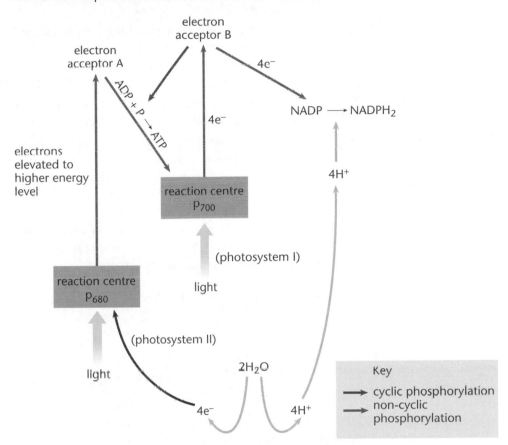

Non-cyclic photophosphorylation

- Light reaches the chlorophyll of both photosystems (P_{680} and P_{700}) which results in the excitation of electrons.
- Electron acceptors receive these electrons (**accepting** electrons is **reduction**).
- P_{680} and P_{700} have become oxidised (**loss** of electrons is **oxidation**).
- P_{680} receives electrons from the **lysis** (splitting) of water molecules and becomes neutral again (referred to as 'hydro lysis').
- Lysis of water molecules releases oxygen which is given off.
- Electrons are elevated to a higher energy level by P_{680} to electron acceptor A and are passed along a series of electron carriers to P_{700}.
- Passage along the electron carrier system funds the production of ATP.
- The electrons pass along a further chain of electron carriers to NADP, which becomes reduced, and at the same time this combines with H^+ ions to form $NADPH_2$.

> After analysing this information you will be aware that in cyclic photophosphorylation P_{700} donates electrons then some are recycled back, hence 'cyclic'. In non cyclic photophosphorylation P_{680} electrons ultimately reach NADP never to return! Neutrality of the chlorophyll of P_{680} is achieved utilising electrons donated from the splitting of water. Different electron sources hence non-cyclic.

Cyclic photophosphorylation

- Electrons from acceptor B move along an electron carrier chain to P_{700}.
- Electron passage along the electron carrier system funds the production of ATP.

Photosynthetic pigments

OCR 4.3.1

Chlorophyll is not just one substance. There are several different chlorophylls, e.g. chlorophyll *a* and chlorophyll *b*.

- Each is a molecule which has a **hydrophilic head** and **hydrophobic tail**.
- The head always contains a **magnesium** ion and plays a key part in the absorbing or harvesting of light.
- The hydrophobic tail anchors to the thylakoid membrane.

> Chlorophyll *a* is the only photosynthetic pigment found in all green plants.

> The role of photosynthetic pigments is to absorb light energy.

As well as the chlorophylls, there are other **accessory pigments**, e.g. carotenoids which also absorb light energy. There are a range of photosynthetic pigments found in different species.

The graphs below show the specific wavelengths of light which are absorbed by a range of pigments. The data for the **absorption spectrum** was collected by measuring the absorption of a range of different wavelengths of light by a solution of each pigment, chlorophyll *a*, chlorophyll *b*, and carotenoids, **separately**. Following this, plants were illuminated at each wavelength of light, in turn, to investigate the amount of photosynthesis achieved at each wavelength. This data is shown in the **action spectrum**.

> The action spectrum shows the actual wavelengths which are used in photosynthesis.

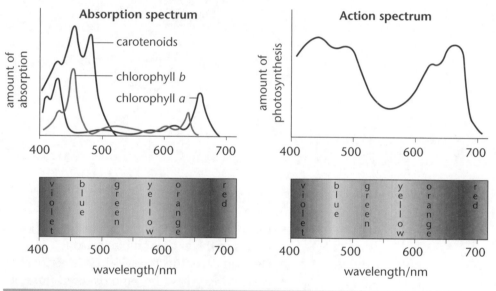

What can be learned from the graphs?

- Blue and red light are absorbed more, and so are key wavelengths for photosynthesis.
- Different pigments have different light absorptive properties.
- Groups of pigments in a chloroplast are therefore much better than just one as more energy can be harnessed for photosynthesis.
- The green part of the spectrum is not absorbed well; no wonder the plants look green as the light is reflected!

Which factors affect photosynthesis?

OCR 4.3.1

If any process is to take place, then correct components and conditions are required. In the case of photosynthesis these are:

- light
- water
- carbon dioxide
- suitable temperature.

Additionally, it is most important that the chloroplasts have been able to develop their photosynthetic pigments in the thylakoid membranes. Without an adequate supply of magnesium and iron, a plant suffers from **chlorosis** due to chlorophyll not developing. The leaf colour becomes yellow-green and photosynthesis is reduced.

Limiting factors

If a component is in low supply then productivity is prevented from reaching maximum. In photosynthesis, **carbon dioxide** is a key limiting factor. The usual atmospheric level of carbon dioxide is 0.04%. In perfect conditions of water availability, light and temperature, this low carbon dioxide level holds back the photosynthetic potential.

Clearly **light energy** is vital to the process of photosynthesis. It is severely limiting at times of partial light conditions, e.g. dawn or dusk.

Water is vital as a photosynthetic component. It is used in many other processes and has a lesser effect as a limiting factor of photosynthesis. In times of water shortage, a plant suffers from a range of problems associated with other processes before a major effect is observed in photosynthesis.

A range of enzymes are involved in photosynthesis; therefore the process has an optimum **temperature** above and below which the rate reduces (so the temperature of the plant's environment can be limiting).

The rate of photosynthesis is limited by light intensity from points A to B. After this, a maximum rate is achieved – the graph levels off.

The rate of photosynthesis is limited by light intensity until each graph levels off. The 30°C graph shows that at 20°C, the temperature was also a limiting factor.

The lower level of CO_2 is also a limiting factor here. The fact that it holds back the process is shown by comparing both graph lines.

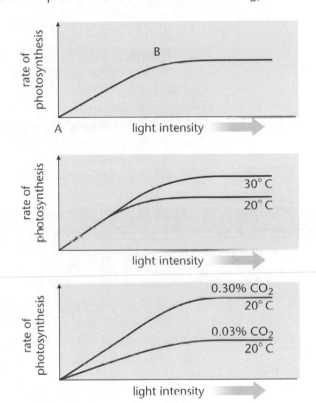

Compensation point

Another way of stating at compensation point is: 'when the rate of respiration equals the rate of photosynthesis'.

It is usual for a plant growing outside in warm conditions to have **two** compensation points every day.

Photosynthesis utilises carbon dioxide whereas respiration results in its production. At night time during darkness, a plant respires and gives out carbon dioxide. Photosynthesis only commences when light becomes available at dawn, if all other conditions are met. At one point, the amount of carbon dioxide released by respiration is totally re-used in photosynthesis. This is the **compensation point**.

Beyond this compensation point, the plant may increasingly photosynthesise as conditions of temperature and light improve. The plant at this stage still respires producing carbon dioxide in its cells and all of this carbon dioxide is utilised. However, much more carbon dioxide is needed which diffuses in from the air.

In the evening when dusk arrives, a point is reached when the rate of photosynthesis falls due to the decrease in light and the onset of darkness. The amount of carbon dioxide produced at one point is totally utilised in photosynthesis. Another compensation point has arrived.

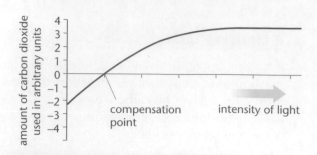

compensation point   intensity of light

How useful is photosynthesis?

Many more substances are synthesised as a result of photosynthesis. Just a few are highlighted in this section.

Without doubt, photosynthesis is a most important process because it supplies carbohydrates and gives off oxygen. There are many more benefits in that glucose is a 'starter' chemical for the synthesis of many other substances. **Cellulose, amino acids** and **lipids** are among the large number of chemicals which can be produced as a result of the initial process of photosynthesis.

The work of the Royal Mint produces the money to run the economy; photosynthesis supplies the **energy currency** for the living world.

The table shows some examples of where and how some carbohydrates are used.

Carbohydrate	Use
deoxyribose (monosaccharide)	DNA 'backbone'
glucose (monosaccharide)	leaves, nectar, blood as energy supply
sucrose (disaccharide)	sugar beet as energy store
lactose (disaccharide)	milk as energy supply
cellulose (disaccharide)	protective cover around all plant cells
starch (polysaccharide)	energy store in plant cells
glycogen (polysaccharide)	energy store in muscle and liver

Progress check

1 In a chloroplast, where do the following take place:
   (a) light-dependent reaction
   (b) light-independent reaction?

2 (a) Which features do photosystems I and II share in a chloroplast?
   (b) Which photosystem is responsible for:
      (i) the elevation of electrons to their highest level
      (ii) acceptance of electrons from the lysis (splitting) of water?

3 (a) Complete the sentence by writing in the correct words.
      The compensation point of a plant is when the rate of ....
      equals the rate of ....

   (b) During a cloudless day in ideal conditions for photosynthesis, how many compensation points does a plant have? Give a reason for your answer.

4 List the three main factors which limit the rate of photosynthesis.

5 During the light-independent stage of photosynthesis, which substances are needed to continue the production of RuBP? Underline the substances in your answer which are directly supplied from the light-dependent stage of photosynthesis.

1 (a) thylakoid membranes (b) stroma.

2 (a) Each photosystem contains a large number of chlorophyll molecules. Light energy is received at the chlorophyll where electrons are boosted to a higher level. Energy is passed to pigment molecules known as the **reaction centre**. The reaction centre of each photosystem absorbs energy (but of different wavelengths).
   (b) (i) photosystem I (ii) photosystem II.

3 (a) respiration; photosynthesis (b) Two. Around dawn and dusk there will come a time when the CO_2 produced as a result of respiration is totally used up in photosynthesis.

4 CO_2; light; temperature.

5 Triose phosphate, NADPH$_2$, ATP and CO_2.

1.3 Respiration

After studying this section you should be able to:

- *recall the structure of mitochondria and relate structure to function*
- *understand that respiration liberates energy from organic molecules*
- *explain the stages of glycolysis and Krebs cycle*
- *explain the stages in the hydrogen carrier system*
- *explain the differences between anaerobic and aerobic respiration*
- *describe different routes which respiratory substrates can take*

The site of respiration

OCR ▸ 4.4.1

Respiration is vital to the activities of every living cell. Like photosynthesis it is a complicated metabolic pathway. The aim of respiration is to break down **respiratory substrates** such as glucose to produce **ATP**.

Respiration consists of a number of different stages. These occur in different parts of the cell. Some stages require oxygen and some do not.

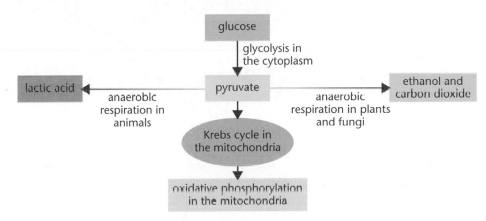

Glycolysis occurs in the cytoplasm of the cell. The pyruvate produced then enters the mitochondria. The Krebs cycle then occurs in the matrix of the mitochondria followed by oxidative phosphorylation which occurs on the inner membrane of the cristae.

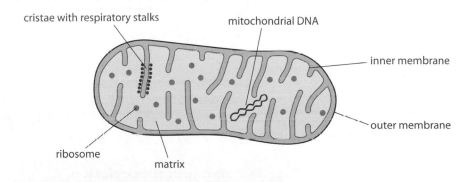

The biochemistry of respiration

OCR ▸ 4.4.1

Glycolysis and the Krebs cycle

Both processes produce ATP from substrates but the Krebs cycle produces **many more** ATP molecules than glycolysis. Every stage in each process is catalysed by a specific enzyme. In aerobic respiration, **both** glycolysis and the Krebs cycle are involved, whereas in anaerobic respiration only glycolysis takes place.

The flow diagram below shows stages in the breakdown of glucose in glycolysis and the Krebs cycle. The flow diagram shows only the main stages of each process.

The two molecules of ATP are needed to begin the process. Each stage is catalysed by an enzyme, e.g. a decarboxylase removes CO_2 from a molecule.

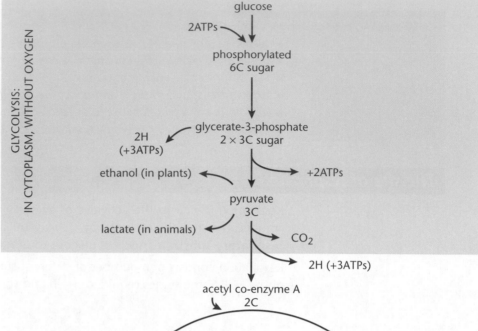

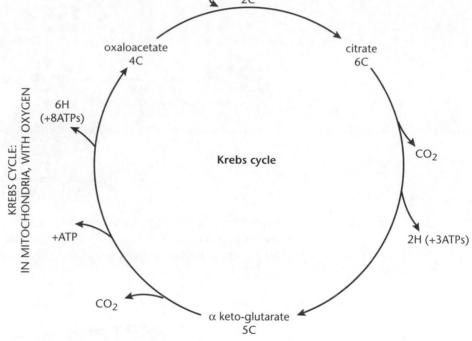

The production of hydrogen atoms during the process can be monitored using DCPIP (dichlorophenol indophenol). It is a hydrogen acceptor and becomes colourless when fully reduced.

The maximum ATP yield per glucose molecule is:

GLYCOLYSIS 2
KREBS CYCLE 2
OXIDATIVE PHOSPHORYLATION 34
= 38 ATP

The flow diagram shows that glycolysis produces $2 \times 2ATP$ molecules but uses 2ATP so the net production is 2ATP. The Krebs cycle makes 2ATP directly. All the rest of the ATP molecules that are made (shown in brackets) are produced in **oxidative phosphorylation**.

Oxidative phosphorylation

The main feature of this process is the electron carrier or electron transport system. The hydrogen that is given off by glycolysis and the Krebs cycle is picked up by acceptor molecules such as **NAD**. These hydrogen atoms are passed along a series of carriers on the inner membrane of the mitochondrion.

Oxygen is needed at the end of the carrier chain as a hydrogen acceptor. This is why we need oxygen to live. Without it, the generation of ATP along this route would be stopped.

Oxidation	Reduction
gain of oxygen	loss of oxygen
loss of hydrogen	gain of hydrogen
loss of electrons	gain of electrons

This is sometimes known as the hydrogen carrier system.

The carrier, NAD, is nicotinamide adenine dinucleotide. Similarly, FAD is flavine adenine dinucleotide.

Hydrogen is not transferred to cytochrome. Instead, the 2H atoms ionise into $2H^+ + 2e^-$. H is passed via an intermediate co-enzyme Q to cytochrome.

Only the electrons are carried via the cytochromes.

e^- is an electron.
H^+ is a hydrogen ion or proton.

An enzyme can be both an oxidoreductase and a dehydrogenase at the same time!

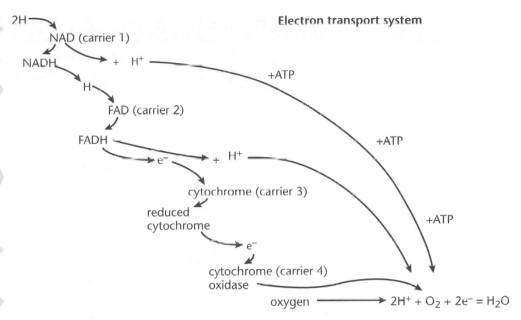

Electron transport system

When oxidation takes place then so does reduction, simultaneously, e.g. $NADH_2$ passes H to FAD. The NAD loses hydrogen and as a result becomes oxidised. FAD gains hydrogen and becomes $FADH_2$, and is therefore reduced. The generic term for an enzyme which catalyses this is **oxidoreductase**. Additionally an enzyme which removes hydrogen from a molecule is a **dehydrogenase**. The result is that three ATPs are produced every time 2H atoms are transported.

The chemiosmotic theory

This theory also explains ATP production in photophosphorylation in the chloroplast. The only difference is that the ions are moved in the opposite direction.

It has now been shown that the carrier molecules are arranged on the membrane of the cristae in a specific way. This means that hydrogen ions are moved out of the matrix and into the space between the two membranes. This sets up a pH gradient. The hydrogen ions can re-enter the matrix through the respiratory stalks. This movement is linked to ATP production and this process is called the **chemiosmotic theory**.

Anaerobic respiration

OCR 4.4.1

If oxygen is in short supply then the final hydrogen acceptor for the hydrogen atoms is missing. This means that oxidative phosphorylation will stop and NAD will not be regenerated. This will result in the Krebs cycle being unable to function.

Ethanol is produced in plants and yeast. Lactate is made in animals.

Glycolysis can continue and produce 2ATP molecules but it would soon run out of NAD as well. A small amount of NAD can be regenerated by converting the pyruvate to lactate or ethanol. This allows glycolysis to continue in the absence of oxygen. This is **anaerobic respiration**.

> Anaerobic respiration will make 2ATP molecules from one glucose molecule compared to a possible 38ATP in aerobic respiration.

KEY POINT

Other respiratory substrates

OCR 4.4.1

For most cells, glucose is the **preferred respiratory substrate**. This means that a cell will respire glucose before using other substances. Once the supply of glucose and glycogen or starch is exhausted, a cell will then respire other respiratory substrates, such as fats and proteins. These substrates enter the pathway in different places.

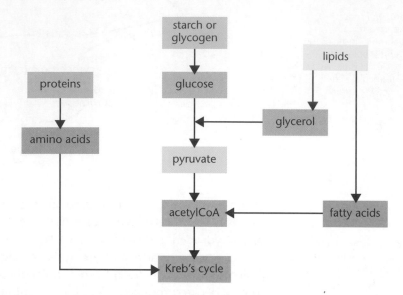

The number of ATP molecules that can be generated from each respiratory substrate depends on the number of hydrogen atoms contained in one mole of the substance. The more hydrogen atoms present, then the more protons that can be released for chemiosmosis. This gives these average energy values for different respiratory substrates:

Respiratory substrate	Energy value in kJ per gram
Carbohydrate	15.8
Lipid	39.4
Protein	17.0

Progress check

1 Explain how hydrogen atom production in cells during aerobic respiration results in the release of energy for cell activity.

2 Give **three** similarities between respiration and photosynthesis.

3 (a) Name the **four** carriers in the electron transport system in a mitochondrion. Give them in the correct sequence.

(b) Name the waste product which results from the final stage of the electron transport system.

4 For each of the following statements indicate whether a molecule would be oxidised or reduced.

(a) (i) loss of oxygen
(ii) gain of hydrogen
(iii) loss of electrons

(b) Which type of enzyme enables hydrogen to be transferred from one molecule to another?

1 Used in the electron transport system to produce ATP; 3ATP molecules produced for every 2H atoms produced; ATP → ADP + P + energy released

2 The stages of each process are catalysed by enzymes; both processes involve ATP; respiration involves GP in glycolysis and photosynthesis involves GP in the light-independent stage

3 (a) NAD → FAD → cytochrome → cytochrome → cytochrome oxidase
(b) water

4 (a) (i) reduced (ii) reduced (iii) oxidised
(b) Oxidoreductase

Sample question and model answer

Radioactivity is used to label molecules. They can then be tracked with a Geiger-Müller counter.

Always be ready to link the rise in one graph line with the dip of another. The relationship holds true here as substance X and RuBP are used up in the production of GP via the Calvin cycle. It is likely that some GP would have been used with substance X to make triose sugar. This is not shown on this graph.

In an experiment, pondweed was immersed in water which was saturated with radioactive carbon dioxide ($^{14}CO_2$). It was illuminated for a time so that photosynthesis took place, the light was then switched off. The graph below shows the relative levels of some substances.

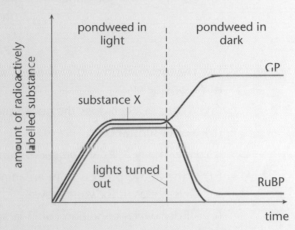

Use the graph and your knowledge to answer the following questions.

(a) (i) Substance X is produced after a substance becomes reduced during the light dependent stage of photosynthesis. Name substance X. [1]

NADPH$_2$, reduced nicotinamide adenine dinucleotide phosphate

(ii) Explain why substance X cannot be produced without light energy. [3]

- Light energy removes electrons from chlorophyll.
- The electrons are passed along the electron carrier chain.
- The electrons are needed to reduce NADP.

(b) Explain the levels of substance X, GP and RuBP after the lights were turned off. [6]

- It seems that substance X is used to make the other two substances because it becomes used up.
- Supply of substance X cannot be produced without light energy.
- GP is made from RuBP.
- GP levels out because more NADPH$_2$ is needed to make triose sugar or RuBP, the supply being exhausted.
- RuBP levels out at a low level because more NADPH$_2$ is needed to make GP.
- ATP is needed to make RuBP, ATP is needed to make GP.

ATP is not shown on the graph. Always be ready to consider substances involved in a process but not shown. Here it is worth a mark to remember that ATP is needed to continue the light-independent system of photosynthesis.

(c) After the lights were switched off glucose was found to decrease rapidly. Explain this decrease. [1]

- Glucose is used up in respiration to release energy for the cell.

(d) Give the specific sites of each of the following stages of photosynthesis in a chloroplast: [2]

(i) light-dependent stage   thylakoid membranes
(ii) light-independent stage.   stroma

Practice examination questions

1 The flow diagram below shows stages in the process of glycolysis.

    2ATPs

glucose → phosphorylated → GP → substance X → lactate
6C sugar   6C sugar   glycerate-   3C
                            3-phosphate
                            (2 × 3C)
                            2ATPs

Use the information in the diagram and your knowledge to answer the questions below.

(a) Where in a cell does the above process take place? [1]

(b) Name substance X. [1]

(c) How many ATPs are *produced* during the above process? [1]

(d) Is the above process from an animal or plant?
    Give a reason for your answer. [1]

(e) Under which condition could lactate be metabolised? [1]

[Total: 5]

2 The graph shows the relative amount of carbon dioxide taken in or evolved by a plant at different times during a day when the sun rose at 5 a.m.

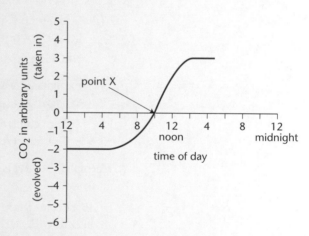

(a) Explain the significance of point X. [2]

(b) What name is given to point X? [1]

(c) Complete the graph between 4.00 p.m. and 12.00 midnight. [2]

[Total: 5]

3 The chlorophyll in a pondweed consisted of several photosynthetic pigments. The graph on the right shows:

(A) the absorption spectrum of the pondweed's chlorophyll *a* measured in arbitrary units

(B) the action spectrum of the same pondweed measured in cm³ oxygen evolved.

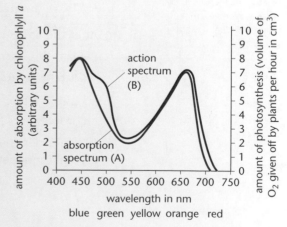

Use the graph and your knowledge to answer the questions.

(a) Explain the similarities and differences between the action and absorption spectra. [2]

(b) Explain the effect of a wavelength of 525 nm on the rate of photosynthesis. [1]

(c) How would the data for the action spectrum have been collected using the pondweed? [1]

[Total: 4]

4 The flow diagram below shows part of the electron carrier system in an animal cell.

$$FADH \rightarrow FAD + H^+ + e^-$$

(a) Where in a cell does this process take place? [1]

(b) From which molecule did FAD receive H to become FADH? [1]

(c) Which molecule receives the electron produced by the breakdown of FADH? [1]

(d) As FADH becomes oxidised a useful substance is produced. Name the substance. [1]

[Total: 4]

5 The graph below shows the effect of increasing light intensity on the rate of photosynthesis of a plant where the concentration of carbon dioxide in the atmosphere was 0.03%.

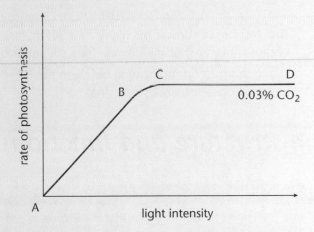

(a) Explain the effect of light intensity on the rate of photosynthesis between the following points on the graph:

(i) A and B
(ii) B and C
(iii) C and D. [3]

(b) Draw the shape of the graph which would result from a CO_2 concentration of 0.3%. [1]

[Total: 4]

Chapter 2
Response to stimuli

The following topics are covered in this chapter:

- Stages in responding to stimuli
- Receptors
- Response

- Neurone structure and function
- Coordination by the CNS
- Plant sensitivity

2.1 Stages in responding to stimuli

After studying this section you should be able to:

- describe the pathway of events that results in response to stimuli

LEARNING SUMMARY

The stimulus/response pathway

OCR 4.1.2, 5.4.2

The ability of plants and animals to respond to changes in their external environment is called **sensitivity**. This is a characteristic of all living organisms and is necessary for their survival. Organisms also respond to changes in their internal environment and this is covered in the next chapter.

The events involved in a response follow a similar pattern:

The responses shown by plants are often less obvious than animal responses because they are usually slower. They still involve a similar pathway of events.

In animals the communication between the receptors, the coordinating centre and the effectors is usually by neurones.

2.2 Neurone structure and function

After studying this section you should be able to:

- describe the structure of a motor neurone, a sensory neurone and a relay neurone.
- understand the function of sensory, motor and relay neurones
- understand nervous transmission by action potential
- describe the mechanisms of synaptic transmission

LEARNING SUMMARY

The structure and function of neurones

OCR 4.1.2

Neurones are **nerve cells** which help to coordinate the activity of an organism by transmitting **electrical impulses**. Many neurones are usually gathered together, enclosed in connective tissue to form **nerves**.

Important features of neurones.

1 Each has a **cell body** which contains a nucleus.
2 Each communicates via processes **from the cell body.**
3 Processes that carry impulses away from the cell body are known as **axons.**
4 Processes that carry impulses towards the cell body are known as **dendrons.**
5 All neurones transmit **electrical impulses.**

The nervous system consists of a range of different neurones which work in a network through the organs. The diagrams show three types of neurone.

> Notice the direction of the impulse and that motor neurones have long axons and short dendrons. This is the other way round for sensory neurones.

Key points from AS

• **The cell surface membrane**
  Revise AS pages 49–50
• **The movement of molecules in and out of cells**
  Revise AS pages 51–52
• **The specialisation of cells**
  Revise AS pages 35–36

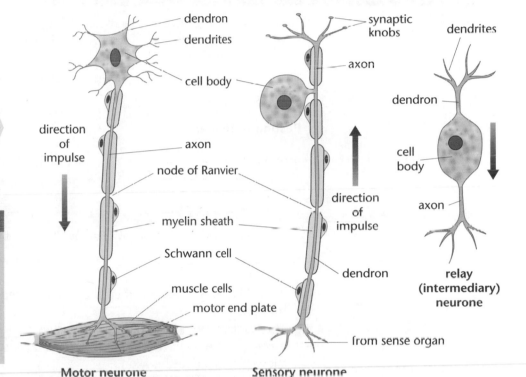

Motor neurone   **Sensory neurone**

Myelinated neurones

Sensory and motor neurones are examples of myelinated neurones. This enables them to transmit an impulse at a greater velocity. Myelinated neurones have the following characteristics:

• The axon or dendron is insulated by a **myelin sheath.**
• The myelin sheath is formed by a **Schwann cell** wrapping around the axon many times. This forms many layers of cell membrane surrounding the axon.
• At intervals there are gaps in the sheath, between each Schwann cell, called **nodes of Ranvier.**

> The myelin sheath is often called a 'fatty' sheath because it is made of many layers of cell membrane which are composed largely of phospholipids.

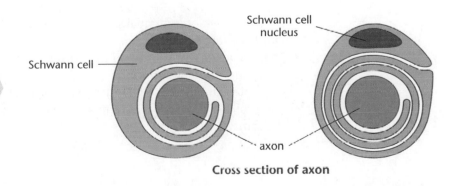

Cross section of axon

What are the roles of the sensory and motor neurones?

There are many similarities between the structure of sensory and motor neurones but they have different functions:

* The sensory neurones transmit impulses towards the central nervous system (CNS) from the receptors.
* The motor neurones transmit impulses from the CNS to effectors, such as muscles, to bring about a response.
* Relay neurones may form connections between sensory and motor neurones in the CNS.

Transmission of an action potential along a neurone

OCR   4.1.2

Neurones can 'transmit an electrical message' along an axon. However, you must never write this in your answers. Instead of nerve impulse, you must now use the term **action potential**.

The diagrams below show the sequence of events which take place along an axon as an action potential passes.

Resting potential

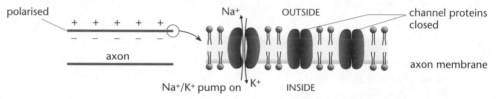

* There are 30 times more Na$^+$ ions on the outside of an axon during a resting potential.
* If any Na$^+$ ions diffuse in, then they are expelled by the '**sodium–potassium pump**'.
* The 'sodium–potassium pump' is an active transport mechanism by which a carrier protein, with ATP, expels Na$^+$ ions against a concentration gradient and allows K$^+$ ions into the axon.
* This creates a **polarisation**, i.e. there is a +ve charge on the outside of the membrane and a –ve charge on the inside.
* The potential difference is called the **resting potential** and can be measured at around –70 millivolts.

Under resting conditions, the membrane of the axon is fairly impermeable to sodium ions.

Action potential – depolarisation

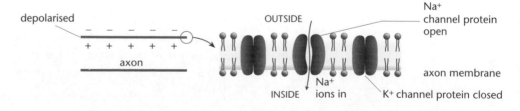

* During an action potential sodium channel proteins open to allow Na$^+$ ions into the axon.
* There is now a –ve charge on the outside and a +ve charge on the inside known as **depolarisation**.
* The potential difference changes to around +50 millivolts.
* The profile of the action potential, shown by an oscilloscope, is always the same.

Action potential – repolarisation

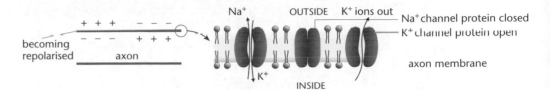

- A K$^+$ channel opens so K$^+$ ions leave the axon.
- This results in the membrane becoming polarised again.
- Any Na$^+$ ions that have entered during the action potential will be removed by the 'sodium–potassium pump'.

Measuring an action potential

- The speed and profile of action potentials can be measured with the help of an oscilloscope.
- The profile of the action potential for an organism always shows the same pattern, like the one shown.
- The changes in potential difference are tracked via a time base.
- Using the time base you can work out the speed at which action potentials pass along an axon as well as how long one lasts.

The diagram shows the typical profile of an action potential.

> All action potentials in one neurone are the same size. A larger stimulus will increase the frequency of action potentials, not the size.

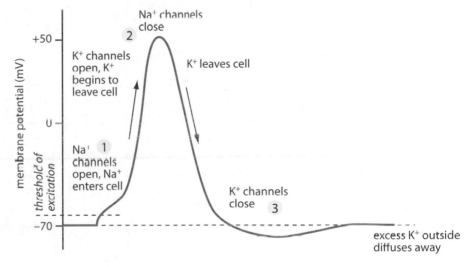

- The front of the action potential is marked by the Na$^+$ channels in the membrane opening.
- The potential difference increases to around +50 millivolts as the Na$^+$ ions stream into the axon.
- The Na$^+$ channels then close and K$^+$ channels in the membrane open.
- K$^+$ ions leave the axon and the membrane repolarises.
- During the **refractory period** no other action potential can pass along the axon, which makes each action potential separate or discrete.

Saltatory conduction

The reason why myelinated neurones are faster than non-myelinated neurones is that the action potential 'jumps' from one node of Ranvier to the next. This is because this is the only place where Na$^+$ ions can pass across the membrane. This is called **saltatory conduction.**

Progress check

What is the function of each of the following?

(a) receptor
(b) axon
(c) myelin sheath
(d) terminal dendrites

(a) Receptors respond to stimulus by producing an action potential.
(b) Transmit action potential with the help of mitochondria.
(c) Myelin sheath is a membrane enclosing fat which acts as an insulator.
(d) Terminal dendrites have motor end plates which can stimulate muscle tissue to contract.

How do neurones communicate with each other?

OCR 4.1.2

The key to links between neurones are structures known as **synapses**. Terminal dendrites branch out from neurones and terminate in **synaptic knobs**. The diagram below shows a synaptic knob separated from an interlinking neurone by a synapse.

Remember an impulse can 'cross' a synapse by chemical means and the route is in ONE direction only. They cannot go back!

A synapse which conducts using acetylcholine is known as a cholinergic synapse.

There are **two** types of synapse:
• **excitatory** which can stimulate an action potential in a linked neurone
• **inhibitory** which can prevent an action potential being generated.

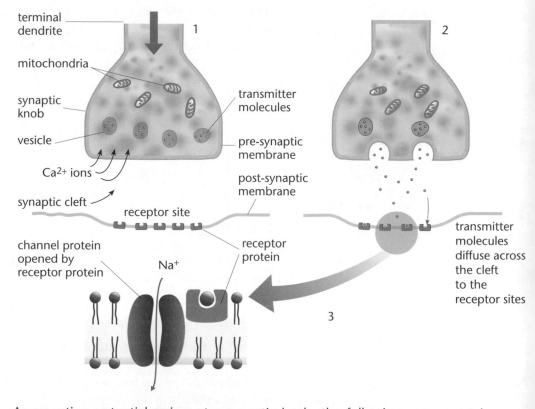

As an action potential arrives at a synaptic knob, the following sequence takes place.

• **Channel proteins** in the **pre-synaptic membrane** open to allow Ca^{2+} ions from the synaptic cleft into the synaptic knob.
• **Vesicles** then merge with the pre-synaptic membrane, so that **transmitter molecules** such as **acetylcholine** are **secreted** into the gap.
• The transmitter molecules diffuse across the cleft and bind with specific **sites** in **receptor proteins** in the **post-synaptic membrane**.
• Every receptor protein then opens a **channel protein** so that ions such as **Na$^+$** pass through the post-synaptic membrane into the cell.
• The Na$^+$ ions **depolarise** the post-synaptic membrane.
• If enough Na$^+$ ions enter then depolarisation reaches a **threshold level** and an **action potential** is generated in the cell.

Response to stimuli

Remember that the generation of an action potential is ALL OR NOTHING. Either enough Na⁺ ions pass through the post-synaptic membrane and an action potential is generated OR not enough reach the other side, and there is no effect.

- Enzymes in the cleft then remove the transmitter substance from the binding sites, e.g. **acetylcholine esterase** removes **acetylcholine** by hydrolysing it into choline and ethanoic acid.
- Breakdown products of transmitter substances are absorbed into the synaptic knob for re-synthesis of transmitter.

KEY POINT

Summation

A single action potential may arrive at a synaptic knob and result in some transmitter molecules being secreted into a cleft. However, there may not be enough to cause an action potential to be generated. If a series of action potentials arrive at the synaptic knob then the build up of transmitter substances may reach the threshold and the neurone will now send an action potential. We say that the neurone has 'fired' as the action potential is produced.

Progress check

1 Explain the importance of summation at a synapse.

2 The diagram shows a synaptic knob.

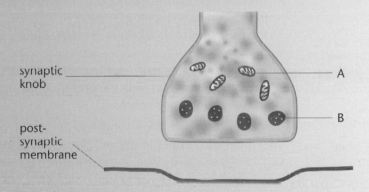

synaptic knob — A

post-synaptic membrane — B

(a) Name A and B

(b) Describe the events which take place after an action potential reaches a synaptic knob and a further action potential is generated as a result.

2 (a) A – mitochondria; B – vesicle
(b) Ca²⁺ ions flow into the synaptic knob; transmitter molecules such as acetylcholine are secreted into the gap; the transmitter molecules bind with sites in receptor proteins in the post-synaptic membrane; this opens channel proteins so that ions such as Na⁺ pass through the post-synaptic membrane into the cell; the post-synaptic membrane is depolarised; if *enough Na⁺ ions enter* a threshold level is reached and an action potential is generated in the cell.

1 A single action potential may arrive at a synaptic knob; there may not be enough transmitter molecules being secreted into a cleft to cause an action potential to be generated; a series of action potentials arrive at the synapse to build up transmitter substances to reach the threshold; the neurone will now send an action potential.

2.3 Receptors

After studying this section you should be able to:

- *list the different types of receptors*

LEARNING SUMMARY

Types of receptors

OCR   4.1.2

The function of receptors is to convert the energy from different stimuli into nerve impulses in sensory neurones.

There are a range of different types of sensory cell around the body. Each type responds to different stimuli. Receptors are classified according to these different stimuli:

The more receptors there are in a position, the more sensitive it is, e.g. the fingers have many more touch receptors than the upper arm.

- **Photoreceptors**, respond to light, e.g. rods and cones in the retina.
- **Chemoreceptors**, respond to chemicals, e.g. taste buds on the tongue.
- **Thermoreceptors**, respond to temperature, e.g. skin thermoreceptors.
- **Mechanoreceptors**, respond to physical deformation, e.g. Pacinian corpuscles in the skin or hair cells in the ear.
- **Proprioreceptors**, respond to change in position in some organs, e.g. in muscles.

Did you know?
The umbilical cord has no receptors. It can be cut without any pain.

Stimulation of a receptor usually causes it to depolarise. This is called a **generator potential**. If this change is beyond a certain magnitude, it will trigger an action potential in a sensory neurone.

Some receptors are found individually in the body such as **Pacinian corpuscles** which detect pressure in the skin. Other receptors are gathered together into sense organs. An example of this is the eye which contains receptors called rods and cones.

2.4 Coordination by the CNS

After studying this section you should be able to:

- *outline the structure and functions of the brain and spinal cord*
- *understand the main functions of cerebrum, cerebellum, medulla oblongata and hypothalamus*

LEARNING SUMMARY

The structure and functions of the CNS

OCR   5.4.2

The CNS consists of the brain and spinal cord which work together to aid the coordination of the organism. The human brain has many functions. The spinal cord takes impulses from the brain to **effectors** and in the opposite direction, impulses from **receptors** are channelled to the brain.

All neurones outside the CNS make up the peripheral nervous system.

The brain has a complex 3D structure. The diagram below shows part of the brain structure – the major components only.

The CNS is like a motorway with impulses going in both directions.

- **Afferent neurones** take impulses **from** organs **to** the CNS.
- **Efferent neurones** take impulses **from the CNS to an organ**.

Learn these carefully. There are no marks for reversal!

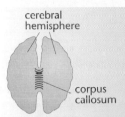

There are two cerebral hemispheres: the left and the right. Note that the right hemisphere controls the left side of the body and vice versa.

Alzheimer's disease

Neurones in the cortex of the cerebrum become progressively less able to produce neurotransmitter substances. Acetylcholine and noradrenaline are usually deficient resulting in major personality changes. The cause is often unknown, but can be genetic.

The hypothalamus is the key structure in maintaining a homeostatic balance in the body. It is like a thermostat in a house, switching the heating system on or off as internal conditions change. Similarly, it is able to control chemical levels in the blood.

The human brain

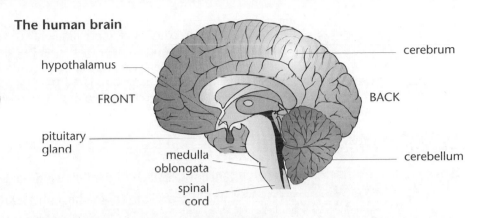

Functions of parts of the brain

Cerebrum

- **Receives sensory information** from many organs, e.g. impulses are sent from the eyes to the visual cortex at the back of the cerebrum.
- **Initiates motor activity** of many organs.
- The front of the cerebrum holds the **memory** and **intelligence** in a network of multi-polar neurones.

Cerebellum

- Has a key role in the coordination of **balance** and smooth, controlled muscular movements.
- Initiation of a movement may be by the cerebrum but the **smooth, well-coordinated** execution of the movement is only possible with the help of the cerebellum.

Medulla oblongata

- Its respiratory centre controls the rhythm of breathing with nerve connections to the intercostal muscles and the diaphragm.
- Its cardiovascular centre controls the cardiac cycle via the sympathetic and vagus nerve.
- Connects to the sino-atrial node of the heart.

Hypothalamus

- Has an exceptional blood supply.
- Many receptors are located in the blood vessel walls which supply it.
- These receptors are highly sensitive detectors which monitor:
  - temperature
  - carbon dioxide
  - ionic concentration of plasma.
- Controls body temperature by various regulatory mechanisms.
- Controls ADH secretion by the pituitary gland and is, therefore, responsible for the water content of both blood plasma and urine.

Pituitary gland

- Secretes a range of hormones and is the major control agent of the endocrine system.
- Responds to neurosecretion and release factors from the hypothalamus.
- Together with the hypothalamus, it is part of a number of negative feedback loops.
- Is the link between the nervous system and the endocrine system.

The above cover some functions of parts of the brain, but there are many more.

Progress check

State two functions of each of the following parts of the human brain:

(a) cerebrum (b) cerebellum (c) medulla oblongata.

(ii) Controls the heart rate via the sympathetic and vagus nerve.

(c) (i) Controls the rhythm of breathing with nerve connections to the intercostal muscles and the diaphragm.

(ii) Enables smooth movement, e.g. hitting a golf ball with a club straight down the fairway. (You could swing the club by voluntary control from the cerebrum but smooth coordination is by the cerebellum.)

(b) (i) Coordinates balance, e.g. enables upright stance in humans.

(ii) Controls voluntary motor activity of the leg muscles.

(a) (i) Receives sensory impulses from the eyes to the visual cortex, enabling sight.

2.5 Response

After studying this section you should be able to:

- describe different types of response
- understand how neurones function together in a reflex arc
- outline the features of the autonomic nervous system
- describe the structure of skeletal muscle and understand the sliding filament mechanism

Different types of response

OCR ▶ 5.4.2

In order to bring about a response, nerve impulses are sent to effectors via motor neurones. Some responses do not require conscious thought. These are called reflexes or reflex actions.

The reflex arc

How can we react quickly without even thinking about making a response?

Often the brain is not involved in the response so the time taken to respond to a stimulus is reduced. This rapid, automatic response is made possible by the **reflex arc**.

A reflex arc

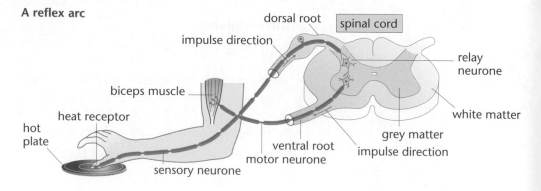

Features of a reflex arc

- The stimulus is detected by a **receptor**.
- As a result, an **action potential** is generated along a sensory neurone.
- The **sensory neurone** enters the spinal cord via the **dorsal root** and **synapses** onto a **relay neurone** / intermediate neurone.
- This intermediate neurone synapses onto a **motor neurone** which in turn conducts the impulse to **a muscle** via its **motor end plates**.
- The muscle contracts and the arm instantly **withdraws** from the stimulus before any harm is done.
- The complete list of events takes place so quickly because the impulses do not, initially, go to the brain! The complete pathway to the muscle conducts the impulse so rapidly, before the brain receives any sensory information.

It is other afferent neurones which *finally* take impulses to the brain enabling us to be aware of the arc which has just taken place. These afferent neurones are NOT part of the reflex arc.

> Reflexes have a high survival value because the organism is able to respond so rapidly. Additionally, they are always automatic. There are a range of different reflexes, e.g. iris/pupil reflex and saliva production.

 KEY POINT

The iris/pupil reflex

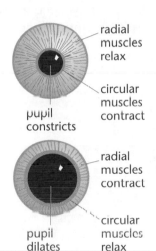

radial muscles relax

circular muscles contract

pupil constricts

radial muscles contract

circular muscles relax

pupil dilates

This response does involve the brain, but because conscious thought is not involved, it is still classed as a reflex.

The diagrams in the margin show the two extremes of pupil size.

- The amount of light entering the eye is detected by **receptors** in the **retina**.
- Reflex pathways lead to the **circular** and **radial** muscles of the **iris**.
- **High-intensity** light activates the **circular muscles** of the iris to **contract**; as the radial muscles relax so the **pupil gets smaller**. (The advantage of this is too much light does not enter and so does not damage the retina.)
- **Low-intensity** light activates the **radial muscles** of the iris to **contract**; as the circular muscles relax so the **pupil gets wider**. (The advantage of this is that the eye allows enough light to see.)
- A balance between the two extremes is achieved across a gradation of light conditions.

Autonomic nervous system

Many reflex actions are controlled by the autonomic nervous system. This is the part of the nervous system which controls our involuntary activities, e.g. the control of the heart rate. It is divided into two parts, the **sympathetic system** and the **parasympathetic system**. Each system has a major nerve from which smaller nerves branch out into key organs. In some ways, the two systems are **antagonistic** to each other but in other ways, they have specific functions not counteracted by the other. The table below shows all of the main facts for each system.

This table of features shows some key points for the autonomic system. ALERT! They are difficult to learn because of the lack of logic in the 'pattern' of functions. Take time to revise this properly because many candidates mix up the features of one system with another.

	Autonomic nervous system	
Feature	*Sympathetic*	*Parasympathetic*
Nerve (example)	sympathetic nerve	vagus nerve
Transmitter substance at synapses	noradrenaline	acetylcholine
Heart rate	speeds up	slows down
Iris control	dilates	constricts
Saliva	————	flow stimulated
Gut movements	slows down	speeds up
Sweating	sweat production stimulated	————
Erector pili muscles	contracts erector pili muscles	————

> Remember that all of the above functions take place without thought. The system is truly involuntary.

Muscles as effectors

OCR ▶ 5.4.2

The body has a number of different effectors, but for most responses the effector is a muscle. There are three types of muscle in the body:

- skeletal/striated or voluntary muscle
- visceral/smooth or involuntary muscle
- cardiac muscle.

Smooth muscle is controlled by the autonomic nervous system, but skeletal muscle is controlled by motor neurones of the somatic nervous system.

How do motor neurones control muscle tissue?

The link to muscle tissue is by **motor end plates** which have close proximity to the sarcoplasm of the muscle tissue. The motor end plates have a greater surface area than a synaptic knob, but their action is very similar to the synaptic transmission described on page 36. Action potentials result in muscle contraction.

No contraction would take place without the acetylcholine transmitter being released from the motor end plate. When the sarcolemma (membrane) reaches the threshold level, then the action potential is conducted throughout the sarcolemma. Contraction is initiated!

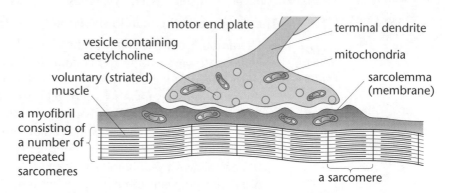

Skeletal muscle is also known as striated or striped muscle. The structure of a single muscle unit, the sarcomere, shows the striped nature of the muscle.

The sarcomere

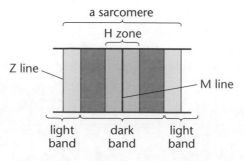

A sarcomere showing bands

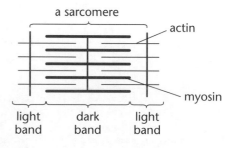

A sarcomere showing filaments

- The sarcomere consists of different filaments, thin ones (**actin**) and thick ones (**myosin**).
- These filaments form bands of different shades:
  – light band (I bands) – just actin filaments
  – dark band (A bands) – just myosin filaments or myosin plus actin.
- During contraction, the filaments slide together to form a shorter sarcomere.
- As this pattern of contraction is repeated through 1000s of sarcomeres, the whole muscle contracts.

- Actin and myosin filaments slide together because of the formation of cross bridges which alternatively build and break during contraction.
- Cross bridge formation is known as the 'ratchet mechanism'.

How does the 'ratchet mechanism' work?

To answer this question, the properties of actin and myosin need to be considered. The diagram below represents an actin filament next to a myosin filament. Many 'bulbous heads' are located along the myosin filaments (just one is shown). Each points towards an actin filament.

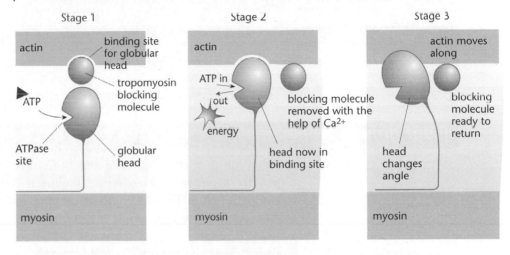

The sequence of the ratchet mechanism

- Once an action potential is generated in the muscle tissue then **Ca²⁺ ions** are released from the **reticulum**, a structure in the **sarcoplasm**.
- Part of the **globular head** of the myosin has an **ATPase** (enzyme) site.
- Ca²⁺ ions activate the myosin head so that this ATPase site hydrolyses an ATP molecule, **releasing energy**.
- Ca²⁺ ions also bind to **troponin** in the actin filaments, this in turn removes **blocking molecules (tropomyosin)** from the actin filament.
- This exposes the **binding sites** on the **actin** filaments.
- The globular heads of the myosin then bind to the newly exposed sites forming **actin–myosin cross bridges**.
- At the stage of energy release the myosin **heads change angle**.
- This change of angle moves the actin filaments towards the centre of each sarcomere and is termed the **power stroke**.
- More ATP binds to the myosin head, effectively causing the cross bridge to become straight and the tropomyosin molecules once again block the actin binding sites.
- The myosin is now 'cocked' and ready to repeat the above process.
- Repeated cross bridge formation and breakage results in a rowing action shortening the sarcomere as the filaments slide past each other.

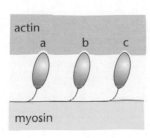
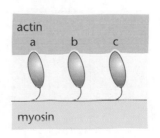
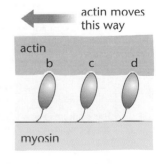

How does skeletal muscle produce movement?

Motor neurones control the skeletal muscle via motor end plates. The skeletal muscles move the bones via their tendon attachments. The muscles work in **antagonistic** pairs, i.e. **opposing** each other. In the arm, when the biceps contracts the forearm is lifted. At the same time the triceps relaxes. If the forearm is to be lowered then the triceps contracts and the biceps relaxes.

2.6 Plant sensitivity

After studying this section you should be able to:

- *understand the range of tropisms which affect plant growth*
- *understand how auxins, gibberellins and cytokinins control plant growth*
- *understand how phytochromes control the onset of flowering in plants*

LEARNING SUMMARY

Plant growth regulators

OCR ▶ 5.4.1

External stimuli such as light can affect the direction of plant growth. A **tropism** is a **growth response** to an external stimulus. It is important that a plant grows in a direction which will enable it to obtain maximum supplies. Plant regulators are substances produced in minute quantities and tend to interact in their effects.

> Growth response to light is **phototropism**
> Growth response to gravity is **geotropism**
> Growth response to water is **hydrotropism**
> Growth response to contact is **thigmotropism**
> Tropisms can be positive (towards) or negative (away from).
>
> **KEY POINT**

Phototropism

This response is dependent upon the stimulus – light affecting the growth regulator, **auxin** (indoleacetic acid).

Auxin and growing shoots

Auxin is produced by cells undergoing mitosis, e.g. growing tips. If a plant shoot is illuminated from one side then the auxin is redistributed to the side furthest from the light. This side grows more strongly, owing to the elongation of the cells, resulting in a bend towards the light. The plant benefits from increased light for photosynthesis. Up to a certain concentration, the degree of bending is proportional to auxin concentration.

stick

auxin high concentration here so cells elongate

Thigmotropism helps a climbing plant like the runner bean to grow in a twisting pattern around a stick. Auxin is redistributed away from the contact point so the outer cells elongate giving a stronger outer growth.

The diagrams show tropic responses to light and auxin.

Tropisms in response to light from different directions

Tropism in response to auxin

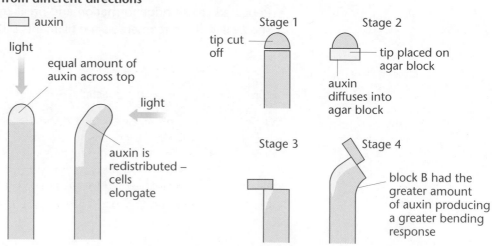

☐ auxin

light

equal amount of auxin across top

light

auxin is redistributed – cells elongate

Stage 1
tip cut off

Stage 2
tip placed on agar block

auxin diffuses into agar block

Stage 3

Stage 4

block B had the greater amount of auxin producing a greater bending response

Auxin research

Many investigations of auxins have taken place using the growing tips of grasses. Where a growing tip is removed and placed on an agar block, auxin will diffuse into the agar. Returning the block to the mitotic area stimulates increased cell elongation to the cells receiving a greater supply of auxin.

Is the concentration of auxin important?

It is important to consider the implications of the concentration of auxin in a tissue. The graph below shows that at different concentrations, auxin affects the shoot and the root in different ways.

Analyse this graph carefully. It shows how the same substance can both stimulate or inhibit, depending on concentration.

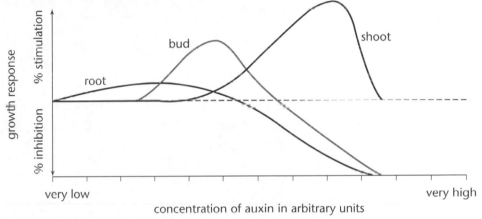

The graph shows:

- auxin has no effect on a shoot at very low concentration
- at these very low concentrations root cell elongation is stimulated
- at higher concentrations the elongation of shoot cells is stimulated
- at these higher concentrations auxin inhibits the elongation of root cells.

Auxin and root growth

The graph above shows that auxin affects root cells in a different way at different concentrations. At the root tip, auxin accumulates at a lower point because of gravity. This inhibits the lower cells from elongating. However, the higher cells at the tip have a low concentration of auxin and do elongate. The net effect is for the stronger upper cell growth to bend the root downwards. The plant therefore has more chance of obtaining more water and mineral ions.

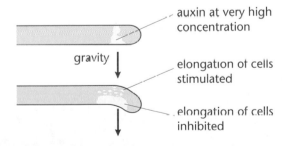

Plant growth regulators

In your examination, look out for data which will be supplied, e.g. the growth regulator gibberellin may be linked to falling starch levels in a seed endosperm and increase in maltose. Gibberellin has stimulated the enzymic activity.

Hormone	Some key functions
auxin	increased cell elongation, suppression of lateral bud development
gibberellin	cell elongation, ends dormancy in buds, promotes germination of seeds by activating hydrolytic enzymes such as amylase (food stores are mobilised!)
cytokinin	increased cell division, increased cell enlargement in leaves inhibits leaf fall
ethene	promotes ripening of food, stimulates leaf fall

Commercial uses of plant growth regulators

Now that many functions of plant growth regulators have been discovered, gardeners and farmers can use them to manipulate the growth of plants.

Growth regulator	Commercial use	Details
Auxins	Taking cuttings	Cuttings are taken from shoots and dipped into rooting powder containing auxins. This stimulates the growth of lateral roots.
	Seedless fruits	If the flowers are treated with auxins, the fruit may develop without fertilisation. It will therefore not contain a seed.
	Weedkillers	Synthetic auxins can be used as selective herbicides (weedkillers). They stimulate rapid growth in the shoots, which then collapse and die.
Gibberillins	Delaying fruit drop	Application of the gibberillins can delay fruit drop, making the fruits riper and preventing damage.
Cytokinins	Tissue culture	Promotes bud and shoot growth in small pieces of plant tissue.
Ethene	Promotes fruit ripening	Fruits can be picked and transported unripe to prevent damage and then ripened with ethene.

Sample question and model answer

(a) The diagram shows a neurone.

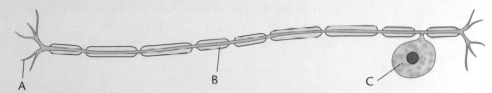

A   B   C

  (i) What type of neurone is shown in the diagram? [1]

      sensory neurone

  (ii) What structure would be found at A? [1]

      a receptor

  (iii) Where precisely in the body is structure C found? [1]

      in the dorsal root ganglion

  (iv) Structure B is covered by a myelin sheath.
Explain the function of the myelin sheath. [3]

      The myelin sheath insulates the neurone;

      Ions can only pass into the neurone at the nodes of Ranvier;

      This speeds up the transmission of the nerve impulse.

(b) The diagram shows a synapse.

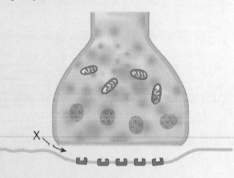

X

  (i) What is contained inside the synaptic vesicles? [1]

      Neurotransmitter / Acetylcholine / transmitter molecules

  (ii) What is the name of the gap marked X? [1]

      Synaptic cleft

  (iii) Why are there so many mitochondria in the end of the neurone? [2]

      To produce ATP;

      For the production of neurotransmitter;

      To reabsorb the neurotransmitter;

      For exocytosis

Practice examination questions

1 The growing tips were removed from oat stems. Agar blocks containing different concentrations of synthetic auxin (IAA) replaced the tips on the oat stems. The plants were allowed to grow for a period then the angle of curvature of the stems was measured. The results are shown in the graph below.

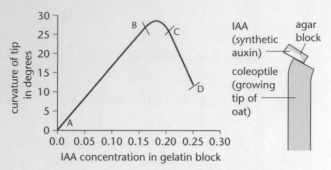

(a) What is the relationship between IAA concentration and curvature of the stem between points:

    (i) A and B   [1]

    (ii) C and D?   [1]

(b) Explain how IAA causes a curvature in the oat stems.   [2]

(c) Explain the effect a much higher concentration of IAA would have on the curvature of oat stems.   [2]

[Total: 6]

2 The diagram below shows a single sarcomere just before contraction.

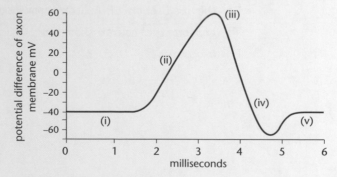

(a) Name filaments A and B.   [2]

(b) What stimulus causes the immediate contraction of a sarcomere?   [1]

(c) What happens to each type of filament during contraction?   [2]

[Total: 5]

3 The diagram below shows the profile of an action potential.

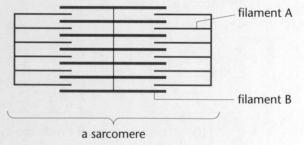

Explain what happens in the axon at each stage shown on the diagram.   [10]

[Total: 10]

Chapter 3
Homeostasis

The following topics are covered in this chapter:

- Hormones
- Regulation of blood sugar level
- Temperature control in a mammal
- The kidneys

3.1 Hormones

After studying this section you should be able to:

- *define homeostasis*
- *describe the route of hormones from source to target organ*
- *understand that hormones and nerves contribute to homeostasis*

LEARNING SUMMARY

The endocrine system

OCR   4.1.1/3

The endocrine system secretes a number of chemicals known as **hormones**. Each hormone is a substance produced by an **endocrine gland**, e.g. adrenal glands produce the hormone adrenaline. Each hormone is **transported in the blood** and has a **target organ**. Once the target organ is reached, the hormone **triggers a response** in the organ. Many hormones do this by **activating enzymes**. Others **activate genes**, e.g. steroids.

> The great advantage of homeostasis is that the conditions in the environment fluctuate but conditions in the organism remain stable.

> The endocrine and nervous systems both contribute to **coordination** in animals. They help to regulate internal processes. **Homeostasis** is the maintenance of a **constant internal environment**. Nerves and hormones have key roles in the maintenance of this **steady internal state**. Levels of pH, blood glucose, oxygen, carbon dioxide and temperature all need to be controlled.
>
> **KEY POINT**

Parts of the human endocrine system (both male and female organs shown!)

> The production of adrenaline prepares the body for 'fight or flight'. It increases the heart rate and breathing rate. It also converts glycogen into glucose.

pituitary gland
thyroid stimulating hormone (TSH)
antidiuretic hormone (ADH)
follicle stimulating hormone (FSH)
luteinising hormone (LH)

hypothalamus (release and inhibitory factors)

thyroid gland (thyroxine)

stomach mucosa (gastrin)

pancreas (insulin and glucagon)

small intestine mucosa (secretin and pancreozymin / cholecystokinin)

adrenal gland (adrenaline)

testis (testosterone)

ovary (oestrogen and progesterone)

49

How does a hormone trigger a cell in a target organ?

When a cell responds to a hormone or a nerve impulse or a chemical, it is called **cell signalling**.

Hormones are much slower in eliciting a response than the nervous system. Rather than having an effect in milliseconds like nerves, hormones take longer. However, effects in response to hormones are often **long lasting**.

The diagram below shows one mechanism by which hormones activate target cells.

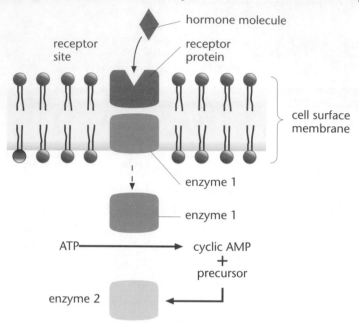

Look carefully at this mechanism! Just **ONE hormone molecule** arriving at the cell releases an enzyme which can be used **many** times. In turn, another enzyme is produced which can be used **many** times. One hormone molecule leads to **amplification**. This is a cascade effect.

Hormones that are proteins work in this way because they are unable to enter the cell. Steroid hormones, (e.g. oestrogen) can pass through the cell membrane as they are lipid soluble.

Nerves and hormones working together

The nervous system and the endocrine system work together in the body to achieve homeostasis. An example of this cooperation is in the control of the heart rate. Cardiac muscle in the heart is myogenic. This means that it will contract without a nerve impulse. The rate is set by the pacemaker in the SA node but this can be adjusted by nerves and hormones. This allows the body to match the output of the heart to the demands of exercise.

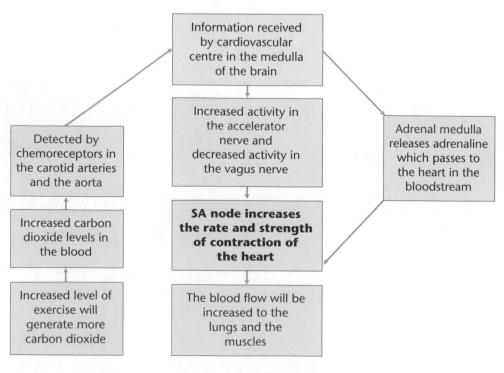

3.2 Temperature control in a mammal

After studying this section you should be able to:

- outline the processes which contribute to temperature regulation in a mammal
- understand how nervous and endocrine systems work together to regulate body temperature
- understand how internal processes are regulated by negative feedback

LEARNING SUMMARY

What are the advantages of controlling body temperature?

OCR ▸ 4.1.1

It is advantageous to maintain a constant body temperature so that the enzymes which drive the processes of life can function at an optimum level.

- **Endothermic** (warm blooded) animals can maintain their core temperature at an optimal level. This allows internal processes to be consistent. The level of activity of an endotherm is likely to fluctuate less than an ectotherm.
- **Ectothermic** (cold blooded) animals have a body temperature which fluctuates with the environmental temperature. As a result there are times when an animal may be vulnerable due to the enzyme-driven reactions being slow. When a crocodile (ectotherm) is in cold conditions, its speed of attack would be slow. When in warm conditions, the attack would be rapid.

Once the blood temperature decreases, the heat gain centre of the hypothalamus is stimulated. This leads to a rise in blood temperature which, in turn, results in the heat loss centre being stimulated. This is negative feedback! The combination of the two, in both directions, contributes to homeostasis.

The **hypothalamus has many functions**. It controls thirst, hunger, sleep and it stimulates the production of many hormones other than those required for temperature regulation.

How is temperature controlled in a mammal?

The key structure in homeostatic control of all body processes is the **hypothalamus**. The regulation of temperature involves thermoreceptors in the skin, body core and blood vessels supplying the brain, which link to the hypothalamus.

The diagram below shows how the peripheral nerves, hypothalamus and pituitary gland integrate nervous and endocrine glands to regulate temperature.

If there is an increase in core temperature then the hypothalamus stimulates greater heat loss by:

- vasodilation (dilation of the skin arterioles)
- relaxing of erector-pili muscles so that hairs lie flat
- more sweating
- behavioural response in humans to change to thinner clothing.

Temperature regulation model

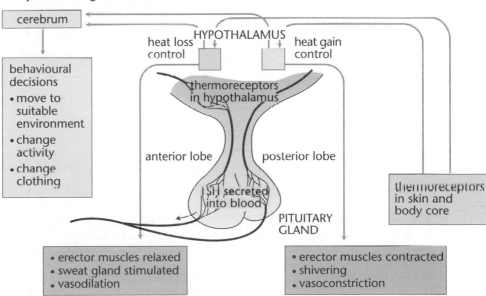

When the hypothalamus receives sensory information **heat loss** or **heat gain** control results.

A capillary bed

arteriole
(a sphincter
muscle)

venule

shunt vessel

artery

vein

A **fall** in temperature results in the following control responses.

Vasoconstriction

- Arteriole control is initiated by the hypothalamus which results in efferent neurones stimulating constriction of the arteriole sphincters of skin capillary beds.
- This deviates blood to the core of the body, so less heat energy is lost from the skin.

Contraction of the erector-pili muscles

- Erector-pili muscle contraction is initiated in the hypothalamus being controlled via efferent neurones.
- Hairs on skin stand on end and trap an insulating layer of air, so less heat energy is lost from the skin.

Sweat reduction

- The sweat glands control is also initiated in the hypothalamus, and is controlled via efferent neurones.
- Less heat energy is lost from the skin by evaporation of sweat.

Shivering

- Increased muscular contraction is accompanied by heat energy release.

Behavioural response

- A link from the hypothalamus to the cerebrum elicits this voluntary response.
- This could be to switch on the heating, put on warmer clothes, etc.

Increased metabolic rate

- The hypothalamus produces a release factor substance.
- This stimulates the anterior part of the pituitary gland to secrete TSH (thyroid stimulating hormone).
- TSH reaches the thyroid via the blood.
- Thyroid gland is stimulated to secrete thyroxine.
- Thyroxine increases respiration in the tissues increasing the body temperature.

Once a higher thyroxine level is detected in the blood the release factor in hypothalamus is inhibited so TSH release by the pituitary gland is prevented. This is **negative feedback**.

Note

(a) the outline for heat loss methods does not show the nerve connections. Efferent neurones are again coordinated via the hypothalamus!

(b) heat is lost from the skin via a combination of **conduction, convection** and **radiation**.

An **increase** in body temperature results in almost the **opposite** of each response described for a fall of temperature.

Vasodilation

- Arterioles of capillary beds dilate allowing more blood to skin capillary beds.

Relaxation of erector-pili muscles

- Hairs lie flat, no insulating layer of air trapped.
- More heat loss of skin.

Sweat increase

- More sweat is excreted so more heat energy from the body is needed to evaporate the sweat, so the organism cools down.

Behavioural response

- This could be to move into the shade or consume a cold drink.

Progress check

Hormone X stimulates the production of a substance in a cell of a target organ. The following statements outline events which result in the production of the substance but are in the wrong order. Write the correct order of letters.

A Hormone X is transported in the blood.
B Hormone X binds with a receptor protein in the cell surface membrane.
C The enzyme catalyses a reaction, forming a product.
D Hormone X is secreted by a gland.
E This releases an enzyme from the cell surface membrane.

D, A, B, E, C.

3.3 Regulation of blood sugar level

After studying this section you should be able to:

- *understand the control of blood glucose levels in a person*
- *describe the sites of insulin and glucagon secretion*
- *explain the functions of insulin and glucagon*

Why is it necessary to control the amount of glucose in the blood?

OCR ▷ 4.2.1, 4.1.3

Glucose molecules are needed to supply energy for every living cell. The level in the blood must be high enough to meet this need (90 mg per 100 cm³ blood). This level needs to be maintained at a constant level, even though a person may or may not have eaten. High levels of glucose in the blood would cause great problems. Hypertonic blood plasma would result in water leaving the tissues by osmosis. Dehydration of organs would result in a number of symptoms.

Remember that the pancreas also produces enzymes. The hormones are released from the islets of Langerhans, which are isolated groups of cells.

Blood glucose regulation

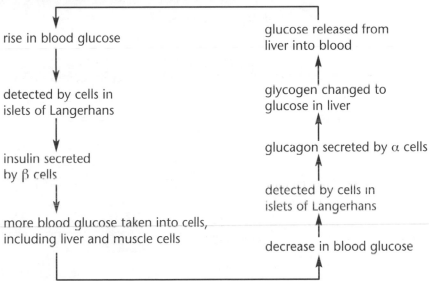

Positive feedback is rare in the body. It means that any change in a variable brings about a response that changes the variable even more.

KEY POINT

Negative feedback
Blood glucose regulation is an example of negative feedback. Any change in glucose level initiates changes which will result in the return of the original level – **balance is achieved.**

Insulin

Never state that insulin changes glucose to glycogen. It allows glucose into the liver where the enzyme glycogen synthase catalyses the conversion.

- Is secreted into the blood due to stimulation of pancreatic cells by a **high concentration** of glucose in the blood.
- Is produced by the β **cells** of the **islets of Langerhans** in the pancreas.
- Binds to receptor proteins in cell surface membranes activating carrier proteins to **allow glucose entry** to cells.
- Allows **excess glucose** molecules into the liver and muscles where they are converted into **glycogen** (a storage product), and some fat.

Glucagon

- Is secreted into the blood due to stimulation of pancreatic cells by a **low concentration** of glucose in the blood.
- Is produced by the α **cells** of the **islets of Langerhans** in the pancreas.
- Stimulates the conversion of glycogen to **glucose.**

Diabetes

There are two types of this condition.

Type 1

- The pancreas fails to produce enough insulin.
- After a meal when blood glucose level increases dramatically, the level remains high.
- High blood glucose causes hyperglycaemia.
- Kidneys, even though they are healthy, cannot reabsorb the glucose, resulting in glucose being in the urine.
- Symptoms include dehydration, loss of weight and lethargy.

What is the answer?

- Insulin injections and a carbohydrate controlled diet.

Type 2

- This form of diabetes usually occurs in later life.
- Insulin is still produced but the receptor proteins on the cell surface membranes may not work correctly.
- Glucose uptake by the cells is erratic.
- Symptoms are similar to those for type 1 but are mild in comparison.

What is the answer?

- Dietary control including low carbohydrate intake.

More liver functions

The role of the liver in its production of bile, as well as the storage and break down of glycogen has been highlighted. The liver does so much more!

Transamination

This is the way an R group of a keto acid is transferred to an amino acid. It replaces the existing R group with another and a new amino acid is formed.

CH$_3$		C$_2$H$_5$		C$_2$H$_5$		CH$_3$	Note the changes in the 'R' group of each acid.
amino acid	+	keto acid	→	amino acid	+	keto acid	

> Children need 10 essential amino acids (adults need 8). From these they can make different ones by transamination in the liver.

Deamination

This process is necessary to lower the level of **excess amino acids**. They are produced when proteins are digested. Nitrogenous materials have a high degree of toxicity, so the level in the blood must be limited.

$$2NH_2-\underset{\underset{H}{|}}{\overset{\overset{R}{|}}{C}}-COOH \;+\; O_2 \;\rightarrow\; 2\underset{\underset{O}{\|}}{\overset{\overset{R}{|}}{C}}-COOH \;+\; 2NH_3$$

amino acid   oxygen   keto acid   ammonia

> The liver has many functions including:
> - detoxification of poisonous substances
> - heat production.

Ornithine cycle

Ammonia is immediately taken up by ornithine to help make a less toxic substance, urea.

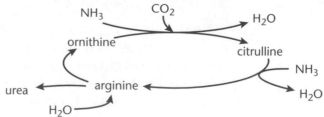

> Both deamination and the ornithine cycle are needed to process excess amino acids. Remember that urea is a less toxic substance. The **liver makes** it but the **kidneys** help to **excrete** it.

3.4 The kidneys

After studying this section you should be able to:

- describe the structure and functions of a nephron
- understand the processes of ultrafiltration and reabsorption
- understand the countercurrent multiplier

LEARNING SUMMARY

Kidney structure and function

OCR   4.2.1

Remember that excretion is the removal from the body of waste products produced by metabolism.

Each kidney has three major regions: the **cortex**, **medulla** and **pelvis**. The renal artery takes blood into a kidney where it is filtered to remove potentially toxic material. Useful substances leave the blood as well as toxic ones but are reabsorbed back into the blood. Toxic substances such as urea leave the kidneys and enter the bladder, via the ureters.

The diagram shows one nephron of the many thousands of nephrons in each kidney.

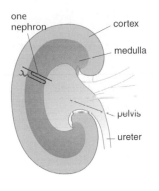

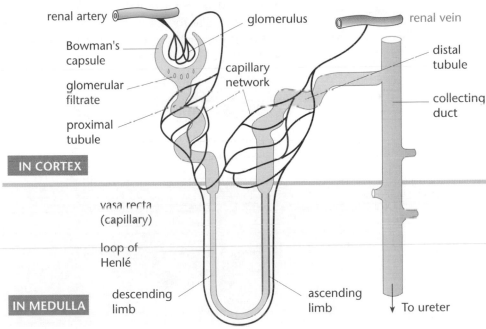

How does a nephron function?

- Blood arrives at the **glomerulus** from the renal artery.
- The **blood pressure is very high** as a result of:
  - contraction of the left ventricle of the heart
  - the arteriole leading to glomerular capillaries is wider than the venule leaving them
  - high resistance of the interface between glomerular capillaries and the inner wall of the renal (Bowman's) capsule.
- **Glomerular filtrate** is forced into the nephron; this is known as **ultrafiltration**.
- Glomerular filtrate includes **urea, glucose, water, amino acids** and **mineral ions**.
- Selective reabsorption takes place in the **proximal tubule** resulting in substances such as glucose being returned to the blood.
- 100% of glucose and 80% of water are reabsorbed at the proximal tubule.
- Urea continues through the tubule to the collecting duct and finally down a ureter to be excreted from the bladder.
- Further reabsorption of substances can take place at the distal tubule.

The selective property of the renal membrane.

What do not leave the blood because they are too large?

Most proteins, red and white blood cells.

Ultrafiltration

Also known as pressure filtration, ultrafiltration relies on the properties of the capillaries and the inner wall of the renal (Bowman's) capsule.

In your examination, you may be requested to label a diagram. Test yourself!

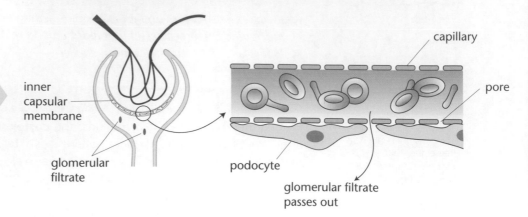

- Capillaries lie very close to the **inner capsular membrane** (see above).
- The capillaries have many tiny pores.
- The capsular membrane consists of **podocytes**.
- **Podocytes** help to support the basement membrane of the capillaries.
- It is the basement membrane that is the selective **high-pressure sieve**.
- Only molecules which are small enough can pass through.

Learn the explanations in the bullet points. Bullet points in this book may resemble your examiner's mark scheme.

Reabsorption

Capillaries from the glomerulus extend to a network across both proximal and distal tubules. The close contact between capillary and tubule is important.

Cross section through proximal tubule

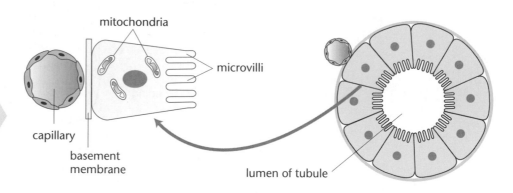

Remember what reabsorption is – the return of substances into the blood which have just left.

- Substances, such as **glucose**, **urea**, and **water** travel along the tubule.
- Each tubule is one cell thick, consisting of epithelial cells with **microvilli** on the outer membrane.
- Microvilli give a **large surface area** to allow the efficient transport of substances to cross to the capillaries.
- **Carrier proteins** on the microvilli **reabsorb** glucose from the filtrate into the tubules.
- Glucose molecules are then actively transported into the fluids surrounding the capillaries.
- Glucose molecules finally enter the capillaries and so have re-entered the blood.
- By the end of the proximal tubule **all glucose** has been **returned to the blood**.

The distal tubule is also in close contact with the capillary network. Even more reabsorption can take place here.

How do the kidneys conserve water?

Water molecules which pass into the tubule and reach the kidney pelvis continue down a ureter and are lost in urine. Such water loss is carefully controlled; some is always reabsorbed. This control involves both the nervous system, the endocrine system and structures along a nephron. The diagram below outlines the role of the **countercurrent multiplier** in the control of water content in the body.

Countercurrent multiplier

The vasa recta capillaries follow the path of the loop of Henlé to:
(a) supply oxygen to the cells so that active transport of Na^+, and Cl^- can take place efficiently (the process needs energy!)
(b) remove CO_2
(c) reabsorb water.

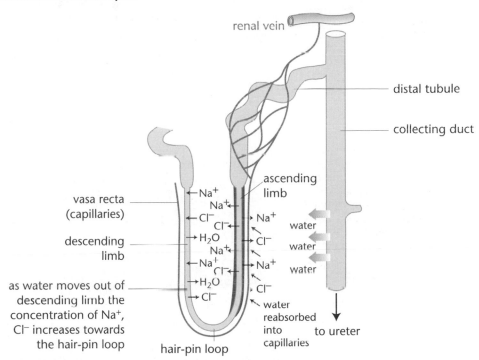

as water moves out of descending limb the concentration of Na^+, Cl^- increases towards the hair-pin loop

The role of the loop of Henlé

- Na^+ and Cl^- ions are actively transported into the medulla from the ascending limb of the loop of Henlé.
- The ascending limb is thicker than the descending one, and impermeable to the outward movement of water so only the ions leave.
- The Na^+ and Cl^- ions slowly diffuse into the descending limb resulting in their greater concentration towards the base of the loop.
- A high concentration of Na^+ and Cl^- ions in the medulla causes water to leave the collecting duct by osmosis.
- Additionally, water leaves the descending limb by osmosis due to the ions in the medulla.
- Water molecules pass into the capillary network and are successfully reabsorbed.

The role of the distal tubule

The distal tubule is also a site of more reabsorption. Even more substances are returned to the blood here.

The structure of the distal tubule is similar to the proximal tubule, however its specific roles are:
- maintenance of a constant blood plasma pH at around 7.4
- if blood plasma falls **below** a pH of 7.4 then ionic movements take place
  - (H^+ ions)   plasma → filtrate
  - (HCO_3^- ions)   filtrate → plasma
- if blood plasma **rises** above a pH of 7.4 then more ion movements take place
  - (OH^- ions)   plasma → filtrate
  - (HCO_3^- ions)   plasma → filtrate

The control of water balance

It is necessary to control the amount of water in the blood. The kidneys can help to achieve this with their ability to intercept water before it can reach the ureters. There are, however, problems to overcome.

In hot conditions we lose a lot of water by sweating; too much loss would lead to dehydration problems.

In cold conditions much less water is lost by sweating, giving a potential problem of too much water being retained in the blood.

A balance must be achieved!

Here the consequences of the two extremes of hot and cold are explained. Do remember that there are a range of conditions **between** these extremes. ADH level changes in response to osmoreceptor sensory input to the hypothalamus.

Hormonal control of the kidneys – the role of ADH

Control is achieved with the help of antidiuretic hormone (ADH), produced by the hypothalamus and secreted by the posterior lobe of the pituitary gland.

Scenario 1: warm environmental conditions
- **Osmoreceptors** in the hypothalamus detect an **increase** in the solute concentration of the blood plasma.
- The **hypothalamus** then produces, by neurosecretion, the hormone **ADH**.
- The ADH is secreted into the posterior lobe of the **pituitary gland**.
- From here it passes into the blood and finally reaches the target organs, the **kidneys**.
- Here ADH **increases** permeability of:
  (i) the collecting ducts
  (ii) the distal tubules.
- The effect is that more water can be **reabsorbed** back into blood.

The events outlined above give a maximum effect of the countercurrent multiplier. Too much water would be lost by sweating so the water component of the urine must be drastically limited. The resulting urine is therefore low in water content and high in solutes.

Look out for graphs in questions about kidneys.
- Levels of key substances may be shown.
- If water content down a collecting duct decreases as water content in the medullary region increases:
  – then water molecules are crossing the collecting duct
  – sodium and chloride ions have drawn this water from the collecting duct into the medulla by osmosis.

Scenario 2: cold environmental conditions
- **Osmoreceptors** in the hypothalamus detect a **decrease** in the solute concentration of the blood plasma.
- The hypothalamus then produces **less ADH**.
- Less ADH leaves the posterior lobe of the pituitary gland.
- Less ADH reaches the target organs, the kidneys.
- The collecting ducts and the distal tubules are **not so permeable**.
- **Less water** can be **reabsorbed** back.

The urine is of greater volume due to greater water content. No wonder we urinate more in cold weather!

Renal dialysis

Anticoagulant is added to stop the blood clotting whilst it is in the machine.

People may have kidney failure for a number of reasons. This may require them to have renal dialysis. This involves linking the person up to a dialysis machine at regular intervals. Blood is removed from a vein in the arm. It then passes through a long coiled tube made of partially permeable cellophane. The fluid surrounding the tube contains water, salts, glucose and amino acids but no waste products such as urea. These waste materials therefore diffuse out of the blood into the fluid.

Sample question and model answer

The diagram shows a cell of the inner wall of a renal (Bowman's) capsule. The two structures shown in the diagram are very important in the passage of substances out of the blood into the proximal tubule.

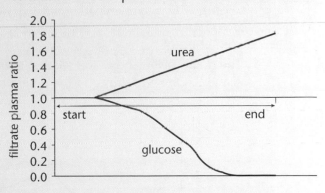

Note the close proximity of cell A to the capillary. This gives a clue as to its function.

(a) Name cell A. [1]

podocyte

(b) Explain how the cells of the inner capsule wall and the capillaries of the glomerulus help in the process of ultrafiltration. [5]

The question shows that five marks are available. Make sure that you give at least five points to gain your marks. Superficial answers fall short of the total.

Capillaries lie very close to the inner capsular wall; the capillaries have pores; the podocytes are shaped so that many gaps exist between the capsular wall and capillaries; the podocytes help to support the capillary basement membrane which is under pressure; only molecules which are small enough are forced through the basement membrane so the process is selective.

(c) As glomerular filtrate leaves the renal (Bowman's) capsule it enters the proximal convoluted tubule. The graph below shows the ratio of glucose and urea in the blood plasma and the filtrate through the proximal tubule. A ratio of 1.0 means that the concentration in both plasma and filtrate are the same.

Explain the changes in plasma filtrate from the beginning to the end of the proximal tubule for:

You may find this difficult. However, you can link the fact that the kidney nephron does reabsorb useful substances but not the waste, urea. Relate this to the graph then your task is possible!

(i) glucose [3]

As the fluid moves along the tubule there is increasingly more glucose in the plasma than the filtrate; this is because glucose is reabsorbed into the blood; all glucose is returned to the blood before end of proximal tubule.

(ii) urea. [3]

As the fluid moves along the tubule there is increasingly more urea in the filtrate than in the blood plasma; no urea is reabsorbed so it remains in the tubule; water is reabsorbed which has the effect of increasing the relative concentration of urea.

Practice examination questions

1 (a) Complete the table below to compare the nervous and endocrine systems. Put a tick in each correct box for the features shown.

	Nervous system	Endocrine system
Usually have longer lasting effects		
Have cells which secrete transmitter molecules		
Cells communicate by substances in the blood plasma		
Use chemicals which bind to receptor sites in cell surface proteins		
Involve the use of Na^+ and K^+ pumps		

[2]

(b) Name the process which keeps the human body temperature and water content of blood regulated. [1]

[Total: 3]

2 A mammal is in hot environmental conditions. Explain the effect of a high quantity of ADH entering the blood from the pituitary gland. [6]

3 (a) The products of transamination are represented below. Complete the equation.

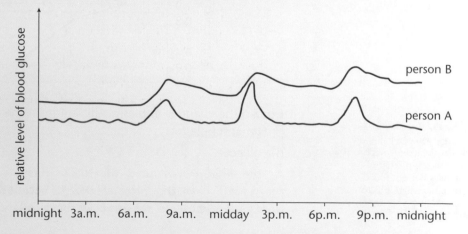

C_2H_5 CH_3

amino acid   keto acid

[2]

(b) (i) Where in the human body does transamination take place? [1]

(ii) Why is transamination necessary in the human body? [2]

[Total: 5]

4 The graph below shows the relative levels of glucose in the blood of two people: A and B. One is healthy and the other one is diabetic.

person B

person A

relative level of blood glucose

midnight 3a.m. 6a.m. 9a.m. midday 3p.m. 6p.m. 9p.m. midnight

(a) Which person is diabetic? Give evidence from the graph for your answer. [1]

(b) What is the evidence that both people produce insulin? [1]

(c) Where in the body is insulin produced? [2]

[Total: 4]

Chapter 4

Further genetics

The following topics are covered in this chapter:

- *Genes, alleles and protein synthesis*
- *Inheritance*
- *Cell division*

4.1 Genes, alleles and protein synthesis

After studying this section you should be able to:

- *explain how proteins are produced in cells*
- *describe various methods by which gene expression is controlled*
- *define a range of important genetic terms*

LEARNING SUMMARY

Genes, alleles and protein synthesis

OCR   S.1.1

A gene is a section of DNA which controls the production of a polypeptide in an organism. The total effects of all of the genes of an organism are responsible for the characteristics of that organism. Each protein contributes to these characteristics whatever its role, e.g. structural, enzymic or hormonal.

The order of bases in the gene is called the genetic code and will code for the order of amino acids in the polypeptide. The order of amino acids is called the primary structure of the protein and will determine how the protein folds up to form the secondary and tertiary structures. The formation of a protein molecule is called protein synthesis.

Protein synthesis

The process of protein synthesis involves the DNA and several other molecules.

- **Messenger RNA:** This is a single-stranded nucleotide chain that is made in the nucleus. It carries the complementary DNA code out of the nucleus to the ribosomes in the cytoplasm.
- **mRNA polymerase:** This is the enzyme that joins the mRNA nucleotides together to form a chain.
- **ATP:** This is needed to provide the energy to make the mRNA molecule and to join the amino acids together.
- **tRNA:** This is a short length of RNA that is shaped rather like a clover leaf. There is one type of tRNA molecule for every different amino acid. The tRNA molecule has three unpaired bases that can bind with mRNA on one end and a binding site for a specific amino acid on the other end.

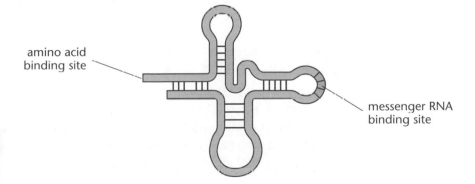

amino acid
binding site

messenger RNA
binding site

Key points from AS

- The genetic code
  Revise AS pages 72–74

The following diagrams show protein synthesis.

1 In the nucleus **RNA polymerase** links to a start code along a DNA strand.

2 RNA polymerase moves along the DNA. For every organic base it meets along the DNA a complementary base is linked to form mRNA (**messenger RNA**).

> There is no thymine in mRNA. Instead there is another base, uracil.

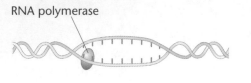

Pairing of organic bases				
DNA	G	C	T	A
mRNA	C	G	A	U

3 RNA polymerase links to a stop code along the DNA and finally the mRNA **moves** to a **ribosome**. The DNA stays in the nucleus for the next time it is needed.

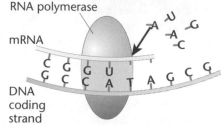

The transfer of the code from DNA to mRNA is called transcription.

4 Every three bases along the mRNA make up one **codon** which codes for a specific amino acid. Three complementary bases form an **anticodon** attached to one end of tRNA (**transfer RNA**). At the other end of the RNA is a specific amino acid.

5 All along the mRNA the tRNA 'partner' molecules enable each amino acid to bond to the next. A chain of amino acids (**polypeptide**) is made, ready for release into the cell.

> Note the link between each pair of amino acids along a polypeptide – the peptide link.

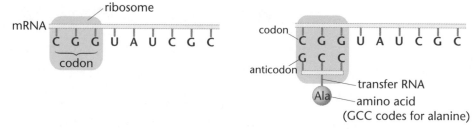

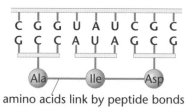

amino acids link by peptide bonds

The conversion of the mRNA code to a sequence of amino acids is called translation.

Control of protein synthesis

OCR 5.1.1

In a multicellular organism, every cell contains all the genetic material needed to make every protein that the organism requires. However, as they develop, cells become specialised. This means that they do not need to use every protein and so it would be a waste to make every protein all the time. Genes must be switched on and switched off.

Most of the early work on gene regulation was carried out on bacteria which are prokaryotic.

Jacob and Monod's theory

During the 1950s, Jacob and Monod found that the bacterium *E.coli* would only produce the enzyme lactase if lactose was present in the growth medium. The production of lactose was controlled by three different genes:

*   a structural gene codes for the enzyme
*   an operator gene which turns the structural gene on
*   a regulator gene that produces a chemical that usually stops the action of the operator gene.

If lactose is present, the action of the chemical inhibitor is blocked and lactase is made.

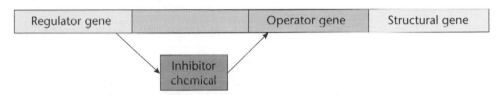

Regulator gene		Operator gene	Structural gene

Inhibitor chemical

> **KEY POINT**
> The combination of the three genes controlling lactase production is called the *lac* operon.

Gene control in eukaryotes

In mature plants, many cells remain totipotent but in mature animals these totipotent or **stem cells** are harder to find.

In eukaryotic cells, gene regulation seems to be much more complicated. Cells in the early embryo are called **totipotent**. This means that they can develop into any type of cell. These cells produce all the cells of a multicellular organism and the specialised cells have to be produced in the correct place.

Scientists are trying to work out how this is done and have found genes called **homeobox genes**. These genes seem to produce proteins that act as transcription factors turning on other genes. Similar homeobox genes have been found in animals, plants and fungi.

Other factors from the cytoplasm can also effect transcription. **Steroid hormones** such as oestrogen can bind with **receptors** in the cytoplasm and then move into the nucleus causing genes to be transcribed.

There is much interest at present in the possible use of siRNA to treat various genetic conditions.

Scientists have recently found a different type of RNA. This is a small double-stranded molecule called siRNA (small interfering RNA). This seems to silence the action of certain genes.

Essential genetic terms

OCR ▶ 5.1.2

It is necessary to understand the following range of specialist terms used in genetics.

Allele – an alternative form of a gene, always located on the same position along a chromosome. This position is called a locus.

> E.g. an allele coding for the white colour of petals

Dominant allele – if an organism has two different alleles then this is the one which is expressed, often represented by a capital letter.

> E.g. an allele coding for the red colour pigment of petals, **R**

Recessive allele – if an organism has two different alleles then this is the one which is **not** expressed, often represented by a lower case letter. Recessive alleles are only expressed when they are not masked by the presence of a dominant allele.

E.g. an allele coding for the white colour pigment of petals, **r**.

Homozygous – refers to the fact that in a diploid organism both alleles for a particular gene are the same.

E.g. **R R** or **r r**.

Heterozygous – refers to the fact that in a diploid organism both alleles for a particular gene are different.

E.g. **R r** (petal colour would be expressed as red).

Co-dominance – refers to the fact that occasionally two alleles for a gene are expressed equally in the organism.

E.g. **A**, **B** alleles = **AB** (blood group with antigens A and B).

Polygenic inheritance – where an inherited feature is controlled by two or more genes, along different loci along a chromosome. This results in continuous variation.

E.g. the height of a person is controlled by a number of different genes.

Haploid – refers to a cell which has a single set of chromosomes.
E.g. a nucleus in a human sperm has 23 single chromosomes.

Diploid – refers to a cell which has two sets of chromosomes.
E.g. a nucleus in a human liver cell has 23 pairs of chromosomes.

Homologous chromosomes – refers to the pairs of chromosomes seen during cell division. These chromosomes lie side by side, each gene at each locus being the same.

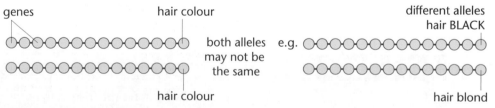

genes   hair colour   both alleles may not be the same   e.g.   different alleles hair BLACK   hair colour   hair blond

Polyploid – refers to the fact that a cell has three or more sets of chromosomes. This can increase yield.

E.g. cultivated potato plants are **tetraploid**, that is four sets of chromosomes in a cell. (Tetraploidy is a form of polyploidy.)

Genotype – refers to all of the genes found in the nuclei of an organism, including both dominant and recessive alleles.

dominant

E.g.   A B c d E F g H i   (all alleles are included in a genotype)
      a B C D e f g h I

recessive

Linkage – refers to two or more genes which are located on the same chromosome.

E.g. linked X-------------------Y--   not linked X-------------------- ◄—different
same chromosomes ╱   -------------------Y-- ◄— chromosomes

Somatic cell – refers to any body cell which is not involved in reproduction.
E.g. liver cell

Autosome – refers to every chromosome apart from the sex chromosomes, X and Y.

4.2 Cell division

After studying this section you should be able to:

- *compare the main features of mitosis and meiosis*
- *describe and explain the process of meiosis*
- *understand the consequences of chiasmata (crossing over)*

LEARNING SUMMARY

Why are there two types of cell division?

OCR   5.1.2

Each type of cell division has a different purpose.

Mitosis

There are occasions when it is necessary to **replicate** cells, e.g. for growth and repair. This is the role of **mitosis**. It produces a clonal line of cells. Each cell divides to form **two diploid** cells, identical in every way.

Meiosis

This is needed in gamete formation. In human cells a body (somatic) cell has 46 chromosomes. If each gamete contained 46 chromosomes then the zygote produced at fertilisation would have 92 chromosomes. This would be lethal! Meiosis is also called **reduction division** because the gametes produced are haploid. In human gametes the haploid chromosome number is 23. Each cell divides to form **4 haploid** cells. Every cell is different to the parent cell and each other.

At AS Level you learned the names of the stages in sequence. The stages of meiosis use the same names, in the same order but there are two nuclear divisions this time!

Meiosis: the process explained

The preparation of a cell prior to meiotic division is during **interphase**. During this pre-stage each double strand of DNA replicates to produce **two** exact copies of itself. This also takes place in exactly the same way before mitosis takes place. After interphase, when the cell division commences, major differences occur.

In meiosis during the first stage, **prophase 1**, a fundamentally important event takes place, where chromatids **cross over**. Each crossover is termed a **chiasma**.

The mechanism of crossovers (chiasmata)

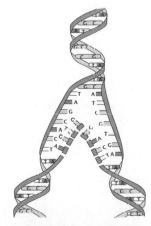

Key points from AS

- Cell division
  Revise AS pages 75–76

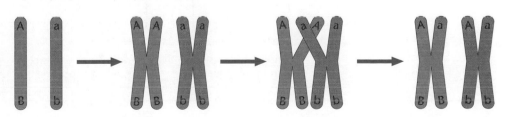

A represents an allele dominant to **a**, a recessive allele.

B represents an allele dominant to **b**, a recessive allele.

In humans, with **many chiasmata** taking place along **all 23 pairs of chromosomes**, every cell at the completion of meiosis is genetically different.

Chiasmata result in **different allele combinations**!

KEY POINT

The process of meiosis

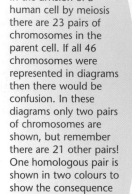

In the division of a human cell by meiosis there are 23 pairs of chromosomes in the parent cell. If all 46 chromosomes were represented in diagrams then there would be confusion. In these diagrams only two pairs of chromosomes are shown, but remember there are 21 other pairs! One homologous pair is shown in two colours to show the consequence of crossovers.

early prophase I

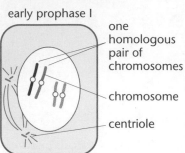

one homologous pair of chromosomes

chromosome

centriole

each chromosome forms two chromatids; two centrioles begin to move forming a spindle

late prophase I

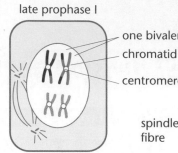

one bivalent

chromatid

centromere

spindle fibre

the chromatids have crossed over and exchanged DNA at two positions

metaphase I

bivalents from homologous chromosomes lie parallel to each other along the equator

anaphase I

corresponding bivalents are pulled by spindle fibres towards poles

telophase I

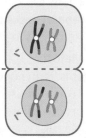

the cell constricts at the equator to form two cells

prophase II

EACH of these cells will divide to form two cells

metaphase II

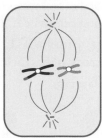

single bivalents lie across the equator

anaphase II

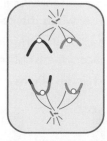

spindle fibres contract to pull the centromere apart; single chromosomes are dragged to the poles

early telophase II

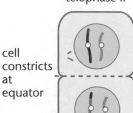

cell constricts at equator

two cells are produced for each cell from first division = four daughter cells

The significance of meiosis

This diagram shows the single chromosomes produced as a result of two crossovers.

In an examination you will need to understand the consequence of many crossovers. Crossovers are a source of genetic variation.

end of first meiotic division

'sister' chromatids still attached

end of second meiotic division

'sister' chromatids have parted from the centromere

The combination of each of these chromosomes with others results in further genetic variation.

A represents an allele dominant to **a**, a recessive allele.

B represents an allele dominant to **b**, a recessive allele.

1 from 4 chromatids combine with 1 from another 4 chromatids. These combinations give **16 possibilities**. Add the combination of another 1 from 4 chromatids and there are **64 possibilities**. Another 1 from 4 is added … and another … to include all 23 pairs. This gives millions of combinations. No wonder we all look different!

- Many more than two crossovers can take place between each homologous pair. The presence of 23 homologous pairs of chromosomes in a diploid human cell results in a lot of crossovers.
- Once the chromatids finally separate in **anaphase II**, each moves with 22 others to a pole to produce a daughter cell.
- After division, four different chromatids are produced from each homologous pair (see above).

> **KEY POINT**
>
> What determines which of the four chromatids of one homologous pair is grouped with chromatids from the other homologous pairs?
>
> The answer is '**chance**' and the combination of the 23 single chromosomes dragged through the cytoplasm by the spindle fibres, is known as **independent assortment**.

Every gamete produced by meiosis is genetically different. However, there are two sexes. This means that in sexual reproduction, the fact that there are two different gametes which combine their alleles in the zygote, gives another major source of variation.

Progress check

The diagram below shows a stage in cell division.

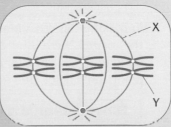

(a) Name parts X and Y.

(b) Which type of cell division is shown? Give a reason for your answer.

(b) Metaphase 1 of meiosis. Bivalents line up in twos along the equator whereas in mitosis they lie singly.

(a) X = spindle fibre   Y = chromatid or bivalent

4.3 Inheritance

After studying this section you should be able to:

- *understand Mendel's laws of inheritance*
- *understand the principles of monohybrid inheritance and dihybrid inheritance*
- *understand exceptions to Mendel's laws such as linkage, sex linkage, co-dominance and epistasis*
- *describe the principle of sex determination*
- *use the Hardy–Weinberg principle to predict the numbers of future genotypes*
- *use chi-squared to test actual genetic data against a predicted ratio*

LEARNING SUMMARY

Mendel and the laws of inheritance

OCR 5.1.2

Gregor Mendel was the monk who gave us our understanding of genetics. He worked with pea plants to work out genetic relationships.

> **KEY POINT**
>
> Mendel's first law indicates that:
> - each character of a diploid organism is controlled by a pair of alleles
> - from this pair of alleles only one can be represented in a gamete.

Monohybrid inheritance

Always show your working out of a genetical relationship in a logical way, just like solving a mathematics problem.

Mendel found that when homozygous pea plants were crossed, a predictable ratio resulted. The cross below shows Mendel's principle.

pea plants   pea plants
T = TALL (dominant)   t = dwarf (recessive)

A homozygous TALL plant was crossed with a homozygous recessive plant

$$TT \quad \times \quad tt$$

gametes   (T) (T)   (t) (t)

F_1 generation   Tt

All offspring 100% TALL, and heterozygous.

If you have to choose the symbols to explain genetics, then use something like **N** and **n**. Here the upper and lower cases are very different. **S** and **s** are corrupted as you write quickly and may be confused by the examiner awarding your marks.

Heterozygous plants were crossed

$$Tt \quad \times \quad Tt$$

gametes   (T) (t)   (T) (t)

	(T)	(t)
(T)	TT	Tt
(t)	Tt	tt

2 × 2 Punnet square to work out different genotypes

F_2 generation   3 TALL : 1 DWARF

In examinations you may have to work out a probability. 3:1 is the same as a 1 in 4 chance. Remember only large numbers would confirm the ratio.

In making this cross Mendel investigated **one** gene only. The height differences of the plants were due to the different alleles. Mendel kept all environmental conditions the same for all seedlings as they developed. The 3:1 ratio of tall to short plants only holds true for large numbers of offspring.

Dihybrid inheritance

Mendel used pea plants to again work out the genetic relationship between plants for genes at different loci (positions) on chromosomes.

> **KEY POINT**
>
> Mendel's second law of independent assortment indicates that:
>
> either of a pair of alleles, say **A** and **a**, can combine with either of another pair, say **B** and **b**.

The cross below shows Mendel's dihybrid principle.

pea plants
R = round seeds (dominant)
Y = yellow seeds (dominant)

pea plants
r = wrinkled seeds (recessive)
y = green seeds (recessive)

A homozygous dominant plant with yellow, round seeds was crossed with a homozygous recessive plant with green, wrinkled seeds.

RRYY × rryy

gametes   (RY)   (ry)

F_1 generation   RrYy

seeds 100% round, yellow, and heterozygous

Below, heterozygous plants from F_1 generation were crossed.

RrYy × RrYy

gametes   (RY) (Ry) (rY) (ry)   (RY) (Ry) (rY) (ry)

F_1 generation

	RY	Ry	rY	ry
RY	RRYY	RRYy	RrYY	RrYy
Ry	RRYy	RRyy	RrYy	Rryy
rY	RrYY	RrYy	rrYY	rrYy
ry	RrYy	Rryy	rrYy	rryy

4 × 4 punnet square to work out genotypes

F_1 generation   9 : 3 : 3 : 1   ratio

round   round   wrinkled   wrinkled
yellow   green   yellow   green

The 9 : 3 : 3 : 1 ratio only holds true for large numbers of offspring.

The principles above can be applied to any dihybrid example. The F_1 generation is so predictable that many varieties of commercial crop are grown from F_1 generation seeds, known as F_1 hybrids.

Sidebar:

You probably will not be given the example of pea plants in your examinations. Coloured petals or fruit fly features are often given. Make sure that you clearly show the symbols and apply the principles learned from this example. The key part in a genetics question is that you give correct gametes.

For a heterozygous genotype of AaBb the gametes are

AB Ab aB ab

NOT

~~AA aa BB bb~~

Be ready to cross two organisms from a punnet square, e.g.

RrYY × rrYY

Do not expect a 9:3:3:1 ratio!

Linkage

OCR   6.1.2

Mendel was rather lucky in the characteristics that he chose. If he had chosen other characteristics there are a number of complications that may have prevented him drawing the correct conclusions. One of these complications is linkage.

Each chromosome consists of a sequence of genes. All genes along a chromosome are **linked** because they are part of the same chromosome. Most chromosomes have between 500 and 1000 genes in a linear sequence. These genes are linked.

Sidebar: Linkage can occur on autosomes and sex chromosomes.

What is the significance of linkage?

We are able to make predictions about the proportion of future offspring when we know the genotype of parents, like the 9:3:3:1 ratio for dihybrid inheritance. This is only true if the pair of contrasting genes are **on different chromosomes**. Consider these two alternatives:

A dominant, **a** recessive; **B** dominant, **b** recessive

      A   a   A   a

loci (positions) of genes | | B   b | |
                  | | B   b

AaBb × AaBb

Not linked This cross would produce 9:3:3:1 proportion in offspring

AaBb × AaBb

Linked This cross would be unlikely to produce a 9:3:3:1 proportion in offspring

69

When two genes, e.g. A a and B b are on **different chromosomes** then their inheritance together is **not affected by crossovers**. Either of one pair **can** be inherited with either of the other pair. The relationship changes when the genes are along the same chromosome. Crossovers are affected! Alleles along the same gene locus can be swapped from one chromatid to another.

Consider these alternatives for linked genes, where a homozygous dominant genotype is crossed with a homozygous recessive.

Crossover 1 (AABB × aabb)

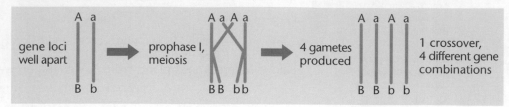

Crossover 2 (AABB × aabb)

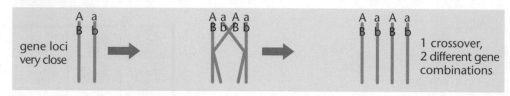

Linkage and probability

Consider these crosses

Cross one

R = red petals (dominant)   r = blue (recessive)
L = long stems (dominant)   l = short stems (recessive)
    homozygous red petals   homozygous blue petals
    long stemmed   short stemmed

	RRLL	×	rrll
gametes	R R		r r
	L L		l l

F$_1$ generation   RrLl   100% heterozygous
                                    red petals, long stemmed

Cross two

	Heterozygous red petals long stemmed		Heterozygous red petals long stemmed

	RrLl	×	RrLl
gametes	R r		R r
	L l		L l

F$_2$ generation

	RL	rl
RL	RRLL red long	RrLl red long
rl	RrLl red long	rrll blue short

probability is: red petals   3:1   blue petals
                         long stems   short stems

actual numbers   610   202

(That is almost the one in four chance.)

This example shows the consequence of **very close linkage**. In this genuine example the genes were so close that the RL and rl combinations were never parted by crossovers. No Rl or rL allele combinations were evident. So the classic RrLl × RrLl ratio of 9:3:3:1 was not possible. Instead a 3:1 ratio was produced. This is **not** monohybrid inheritance.

Progress check

(a) List the gametes for the following dihybrid cross.
    (The genes are not linked.)
        Ddee × DDEe
(b) Show the genotypes of the progeny.

(b) DDEe, DdEe, DDee, Ddee
(a) De de   DE De

Sex determination and sex linkage

OCR ▸ 5.1.2

The genetic information for gender is carried on specific chromosomes. In humans there are 22 pairs of autosomes plus the special sex determining pair, either **XY** (male) or **XX** (female). In some organisms such as birds this is reversed.

Some genes for sex determination are on autosomes but are activated by genes on the sex chromosomes.

Sperm can carry an X or Y chromosome, whereas an egg carries only an X chromosome.

genotype	XX	×	XY
	female		male

gametes   Ⓧ Ⓧ   Ⓧ Ⓨ
                    Ⓧ Ⓧ

offspring		ⓧ	ⓧ
Ⓧ		XX	XX
Ⓨ		XY	XY

probability   1 : 1
                male   female

> The genetic cross shown should not give you any problems at A2 Level. However, look out for the combination of another factor which will increase difficulty.

> Remember that X and Y are chromosomes and not genes.

This shows how 50:50 males to females are produced.

Sex linkage

Look more closely at the structure of the X and Y chromosomes.

> Why do they not look like X and Y? Only when the cells are dividing, do they take the XY shape, after chromatid formation.

X chromosome — non-homologous part of X — homologous part of both X and Y — Y chromosome

Homologous part of the sex chromosomes

* Has the same genes in both sexes.
* Each gene can be represented by the same or different alleles at each locus.

Non-homologous part of the sex chromosomes

* This means that the X chromosome has genes in this area, whereas the Y chromosome, being shorter, has no corresponding genes.
* Genes in this area of the X chromosome are always expressed, because there is no potential of a dominant allele to mask them.

- There are some notable genes found on the non-homologous part, e.g. haemophilia trait, and colour blindness trait.

The sex chromosomes, X and Y, carry genes other than those involved in sex determination. Examples of such genes are:

- a gene which controls blood clotting, i.e. is responsible for the production of factor VIII vital in the clotting process.
- a gene which controls the ability to detect red and green colours

The loci of both genes are on the non-homologous part of chromosome X.

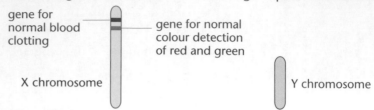

What is the effect of sex linked genes?

The fact that these genes are linked to the X chromosome has no significant effect when the **genes perform their functions correctly**. There are consequences, however, if the genes fail. This can be illustrated by a consideration of **red–green colour blindness**. When a gene is carried on a sex chromosome, the usual way to show this is by X^R.

R = normal colour vision (dominant)   r = red–green colour blindness (recessive)

> You can see the four possible genotypes. A female needs two r alleles, a male just needs one.

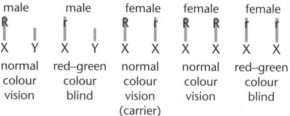

The genetic diagram shows that a female needs two recessive alleles (one from each parent) to be colour blind. A male has only one gene at this locus, so one recessive allele is enough to give colour blindness. The colour blindness gene is rare, so the chances of being a colour blind female are very low.

Consider these crosses.

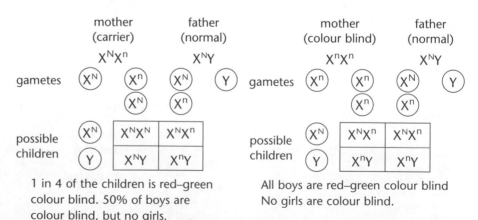

What is the probability of a colour blind male and carrier female producing:
(a) a boy with normal colour vision
(b) a colour blind girl?
Show your working.

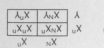

(a) 1 in 4 (b) 1 in 4

Co-dominance

OCR 5.1.2

This term is given when each of two *different* alleles of a gene are expressed in the phenotype of an organism. In humans there are two co-dominant alleles. These alleles produce the antigens in blood which are responsible for our blood groups.

> Remember that for co-dominance there is no dominance. Both alleles are equally expressed.

> Look out for more examples of co-dominance in examination questions, e.g. in shorthorn cattle, R = red and W = white. Where they occur in the phenotype together they produce a dappled intermediary colour known as roan.

Consider these crosses

Blood group	Genotypes
A	$I^A I^A$, $I^A I^O$
B	$I^B I^B$, $I^B I^O$
AB	$I^A I^B$
O	$I^O I^O$

The allele for production of:

A antigen in blood = I^A

B antigen in blood = I^B

No antigen in blood = I^O

Mother × Father
$I^B I^O$ $I^A I^O$

gametes I^B I^O I^A I^O

children	I^B	I^O
I^A	$I^A I^B$	$I^A I^O$
I^O	$I^B I^O$	$I^O I^O$

All blood groups produced by this cross.

Mother × Father
$I^A I^B$ $I^A I^O$

gametes I^A I^B I^A I^O

children	I^A	I^B
I^A	$I^A I^A$	$I^A I^B$
I^O	$I^A I^O$	$I^B I^O$

There must be a I^O from both parents to produce an O blood group.

In this instance, there are two co-dominant alleles, I^A and I^B. When inherited together they are both expressed in the phenotype. Group O blood does not have any antigen.

Epistasis

OCR 5.1.2

This involves two different genes which affect each other. A form of epistasis can be explained by referring to the sweet pea plant. *Lathyrus odoratus* is a white flowered sweet pea. When crossed, two white parent plants can produce white and purple flowers. This can be explained as follows:

> **Dominant epistasis** also exists. In this instance a dominant allele can **inhibit** another, e.g. in the land snail: **A a** dominant allele inhibits **B, b** alleles responsible for banding on the shell.

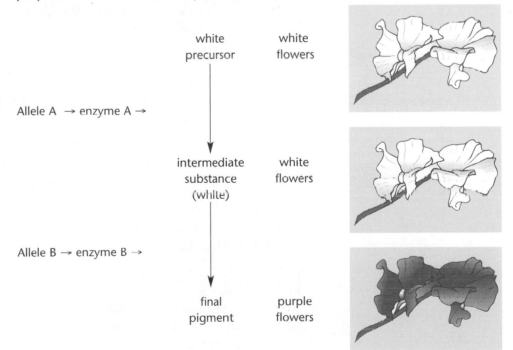

white precursor — white flowers

Allele A → enzyme A →

intermediate substance (white) — white flowers

Allele B → enzyme B →

final pigment — purple flowers

Both alleles A and B are needed to code for their respective enzymes if purple sweet pea flowers are to be produced. The alleles to consider are:

A (dominant)   a (recessive)   B (dominant)   b (recessive)

Genotypes	aabb	aaBb	aaBB	Aabb	AaBB	AaBb	AAbb	AABb	AABB
Phenotypes	white	white	white	white	purple	purple	white	purple	purple

Without the combined effects of both A and B alleles, then the flowers are white. The reliance of one gene on another is an example of **epistasis**.

How is it possible for two white flowered plants to be crossed to give purple offspring?

<table>
<tr><td></td><td>white</td><td></td><td>white</td></tr>
<tr><td>genotype</td><td>AAbb</td><td>×</td><td>aaBB</td></tr>
<tr><td>gametes</td><td>Ab   Ab</td><td></td><td>aB   aB</td></tr>
<tr><td>F₁ generation</td><td colspan="3">AaBb</td></tr>
</table>

100% purple flowered plants from white flowered parents.

Check out the above genotypes to find two more genotypes of white flowered plants which could be crossed to give purple offspring.

> Questions about epistasis usually give some data which you will need to analyse. The organisms may not be sweet pea plants but the principles remain the same.

Hardy–Weinberg principle

OCR   5.1.2

The application of this principle allows us to **predict numbers of expected genotypes** in a population in the future. The principle tracks the proportion of two different alleles in the population.

Before applying the Hardy–Weinberg principle, the following criteria must be satisfied.

- There must be no immigration and no emigration.
- There must be no mutations.
- There must be no selection (natural or artificial).
- There must be true random mating.
- All genotypes must be equally fertile.

Once the above criteria are satisfied then **gene frequencies remain constant**.

> A **gene pool** consists of all genes and their alleles, which are part of the reproductive cells of an organism. Only genes that are in cells that **can be passed on** are part of the gene pool.

Hardy–Weinberg principle: the terms identified

p = the frequency of the dominant allele in the population

q = the frequency of the recessive allele in the population

p^2 = the frequency of homozygous dominant individuals

q^2 = the frequency of homozygous recessive individuals

$2pq$ = the frequency of heterozygous individuals

> **KEY POINT**
>
> The principle is based on two equations:
> (i) $p + q = 1$ (gene pool)
> (ii) $p^2 + 2pq + q^2 = 1$ (total population)

Applying the Hardy–Weinberg principle

A population of *Cepaea nemoralis* (land snail) lived in a field. In a survey there were 1400 pink-shelled snails and 600 were yellow. There were two alleles for shell colour.

Y = pink shell (dominant) y = yellow shell (recessive). Snails with pink shells can be YY or Yy. Snails with yellow shells can be yy only.

phenotype	pink	yellow
genotype	YY Yy	yy

This part of the calculation is to find the frequency of the recessive and dominant alleles in the population.

$$q^2 = \frac{600}{2000}$$
$$= 0.30$$
$$q = \sqrt{0.30} = 0.55$$

But: $p + q = 1$
$$p = 1 - 0.55$$
$$= 0.45$$

So: $p^2 = 0.20$

This part of the calculation is to find the frequency of homozygous and heterozygous snails in the population.

But: $p^2 + 2pq + q^2 = 1$
$$0.20 + 0.50 + 0.30 = 1$$
$$\text{YY} \quad \text{Yy} \quad \text{yy}$$

Points to note

- These proportions can be applied to the snail populations in say, 10 years in the future.
- If there were 24 000 snails in the population, then the relative numbers would be:
  YY $0.2 \times 24\,000 = 4800$
  Yy $0.5 \times 24\,000 = 12\,000$
  yy $0.3 \times 24\,000 = 7200$
- Remember that the five criteria must be satisfied if the relationship is to hold true.
- It is not possible to see which snails are homozygous dominant and which are heterozygous. They all look the same, pink! Hardy–Weinberg informs us, statistically, of those proportions.

It is also possible to apply the Hardy–Weinberg principle to a co-dominant pair of alleles. P and q are calculated by exactly the same method.

Always use the $p + q = 1$ equation to calculate the frequency of alleles if you are given suitable data, e.g. 'out of 400 diploid organisms in a population there were 40 homozygous recessive individuals'. 40 organisms have 80 recessive alleles.

$$q = \frac{80}{800}$$
$$= 0.1$$

From this figure you can calculate the others.

Chi-squared: a statistical test

OCR 5.1.2

When doing scientific investigations, we need to know if our results are significant or due to chance. We should not, for example, conclude that a new genetic ratio we have found represents a significant pattern for a particular cross. The χ^2 (**chi-squared**) test helps us to check out the difference between **expected** results and **actual** results. We can then state the probability that any differences between expected and actual results are due to chance or have significance.

> Remember, in **co-dominance** both alleles are expressed in the phenotype.

$$\chi^2 = \sum \frac{d^2}{x}$$

d = difference between actual and expected results
x = expected results
Σ = the sum of

Consider this example

Dianthus (campion) has flowers of three different colours, red, pink and white. Two pink flowered plants were crossed and the collected seeds grown to the flowering stage.

R = red   r = white (both alleles are co-dominant)

genotypes   Rr × Rr

gametes   R   r   R   r

F_1 generation

	R	r
R	RR	Rr
r	Rr	rr

white 0.25
pink 0.5
red 0.25

> In an examination, you may be given another term for 'actual'. It may be 'observed', but it means the same!

Numbers	RR = red flowers	Rr = pink flowers	rr = white flowers
Actual	34	84	42
Expected	40	80	40

$$\chi^2 = \frac{(40-34)^2}{40} + \frac{(80-84)^2}{80} + \frac{(40-42)^2}{40}$$

$$= \quad 0.9 \quad + \quad 0.2 \quad + \quad 0.1$$

$$= \quad 1.2$$

The next stage is to assess the degrees of freedom for this investigation. This value is always one less than the number of classes of results. In this case there are three classes, i.e. red, pink and white.

Degrees of freedom = (3 − 1) = 2

Now check the χ^2 value against the table.

Degrees of freedom				χ^2				
1	0.00	0.10	0.45	1.32	2.71	3.84	5.41	6.64
2	0.02	0.58	1.39	2.77	4.61	5.99	7.82	9.21
Probability that deviation is due to chance alone (significance level)	0.99 (99%)	0.75 (75%)	0.50 (50%)	0.25 (25%)	0.10 (10%)	0.05 (5%)	0.02 (2%)	0.01 (1%)

> If you are given a χ^2 question in an examination you will be given a data table. A mark may be given for degrees of freedom. Remember, **10** classes of results would give **9** degrees of freedom.

What do you do with the χ^2 value?

- Go to the 0.05 level of significance (5%).
- At 2 degrees of freedom, is the χ^2 value (in this case 1.2) greater than the value given in the table (5.99)?
- If it is greater than there is a significant difference between the observed and expected data.

> In Biology the 0.05 level is the one that is generally used.

Sample question and model answer

(a) Explain the difference between sex linkage and autosomal linkage.　　[2]

　　sex linkage – genes are located on a sex chromosome

　　autosomal linkage – genes are located on one of the other 44 chromosomes

(b) The diagram below shows part of a family tree where some of the people have haemophilia.

This type of question is a challenge! Note the key for the symbols and then apply them to the family tree. Think logically and work up and down the diagram. In your 'live' examination write on the diagram to help you work out each individual genotype asked in the question. If there are a range of possible genotypes they may be helpful.

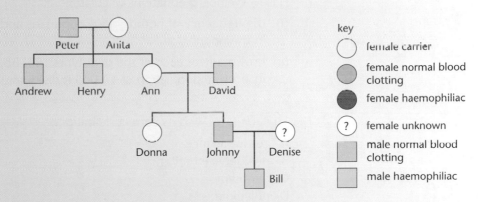

Show the possible genotypes of Denise. Give evidence from the genetic diagram to support your answer.　　[3]

Let H = normal blood clotting

Let h = haemophiliac trait

The genotype can be $X^H X^h$ or $X^h X^h$

Reason – Johnny is $X^H Y$ so he is responsible for Bill's Y chromosome. (Y chromosomes do not carry a blood clotting gene)

Working backwards, Bill is haemophiliac so Denise must have at least one X^h

She can, therefore, be $X^h X^h$ or $X^H X^h$

(c) Peter and Anita had three children. Andrew was born first, then Henry and finally Ann. Use the information in the diagram to answer the questions.

　(i) When could genetic counselling have been given to help Peter and Anita?　[1]

　　After the birth of Henry.

　(ii) Explain the useful information which they could have been given.　　[3]

　　Since Henry is haemophiliac his genotype is $X^h Y$.

　　His father, Peter, has normal clotting blood so is $X^H Y$ and passes on a Y to his son, Henry.

　　His mother is not haemophiliac but must be a carrier, $X^H X^h$ because mother passes on X^h.

　　We can predict 2 in 4 children will have normal clotting of blood, 1 in 4 will be female and a carrier and 1 in 4 will be haemophiliac male.

Practice examination questions

1 (a) List the criteria which must be satisfied before applying the Hardy–Weinberg principle. [4]

(b) In a population of 160 small mammals, some had a dark brown coat and the others had a light brown coat. Dark brown (B) is dominant over light brown (b). In the population there were 48 light brown individuals. Using the Hardy–Weinberg equations calculate:

(i) the frequency of homozygous dominant and heterozygous individuals in the population [3]

(ii) how many of each of the genotypes (BB, Bb, bb) there would be in a future population of 10 000 individuals. [2]

[Total: 9]

2 Match each term with its correct definition.

A co-dominance
B polygenic inheritance
C genotype
D polyploid
E somatic

(i) a cell which is not involved in reproduction [1]

(ii) a nucleus which has three or more sets of chromosomes [1]

(iii) a feature which is controlled by two or more genes, along different loci along a chromosome [1]

(iv) two alleles which are equally expressed in the organism [1]

(v) all of the genes found in a nucleus, including both dominant and recessive alleles [1]

[Total: 5]

3 The letters below represent the organic bases along the coding strand of a DNA molecule.

CCG ATT CGA TAG

(a) What term is given to each group of three bases? [1]

(b) Give **two** functions of a group of three organic bases. [2]

[Total: 3]

4 The diagram on the right shows a cell at the beginning of telophase II during meiosis.

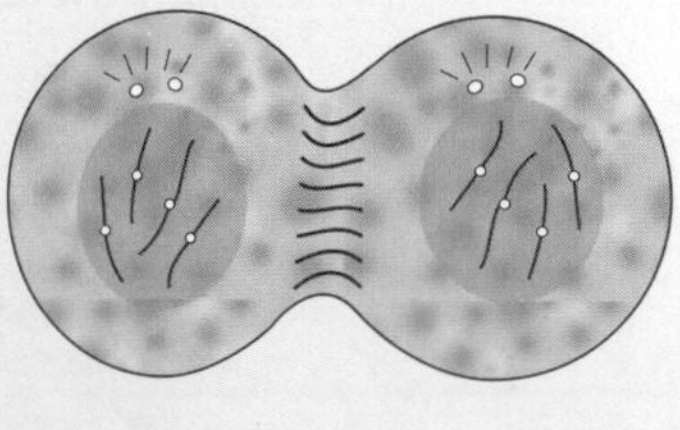

(a) How many chromosomes were there in the parent cell at the beginning of meiosis? [1]

(b) Describe **one** difference between telophase II and:

(i) telophase I of meiosis

(ii) telophase of mitosis. [2]

(c) Describe the stage immediately before telophase II. [2]

[Total: 5]

Chapter 5

Variation and selection

The following topics are covered in this chapter:

- Variation
- Selection and speciation

5.1 Variation

After studying this section you should be able to:

- explain the different sources of variation in organisms
- describe different types of mutation

Variation and mutations

OCR 5.1.1

Meiosis and sexual reproduction can produce variation in a number of ways. These include:

- segregation or independent assortment of homologous chromosomes
- chiasmata formation leading to crossing over
- random fusion of gametes.

All these processes will combine alleles in different combinations.

The environment will also contribute to variation. The combination of environmental variation and a number of genes controlling a characteristic (polygenic inheritance) will often result in a wide range of phenotypes and continuous variation. However, the only way that new alleles can be made is by mutation.

Mutation is a change in the DNA of a cell. If the cell affected by mutation is a **somatic cell**, then its effect is **restricted** to the organism itself. If, however, the mutation affects **gametes**, then the genetic change will be inherited by the future population.

Gene mutations

A gene mutation involves a change in a single gene. This is often a point mutation.

> Bases can change along DNA and this may cause mutation. One changed base along the coding strand of DNA may have a sequential effect of changing most amino acids along a polypeptide.
>
> *before mutation*
> TTA CCG GCC ATC
> *after mutation*
> ATT ACC GGC CAT C
> This is addition!

DNA codes for the sequence of amino acids along polypeptides and ultimately the characteristics of an organism. Each amino acid is coded for by a triplet of bases along the coding strand of DNA, e.g. TTA codes for threonine. The change in a triplet base code can result in a new amino acid, e.g. ATT codes for serine. This type of DNA change along a chromosome is known as a **point mutation**. A point mutation involves a change in a single base along a chromosome by **addition** (insertion), **deletion** or **inversion**.

An example of a gene mutation is a change in the DNA coding for the protein haemoglobin. This can cause sickle-cell anaemia.

Sometimes a point mutation may not cause a change in the phenotype. This is because the genetic code is degenerate. Often one amino acid has more than one triplet coding for it. Therefore a change in a base may not change the amino acid.

> **Key points from AS**
>
> - **Variation**
>   *Revise AS pages 83–84*

More mutations are shown below. Each section of DNA along the chromosomes is shown by organic bases.

Addition

before
TTA CCG GCC ATC

after
CCG TTA CCG GCC ATC

A new triplet has been added. If a triplet is repeated it is also duplication.

Deletion

before
TTA CCG GCC ATC

after
TTA CCG GCC

Inversion

before
TTA CCG GCC ATC

after
TTA CCG GCC **CTA**

CTA codes for a new amino acid.

Translocation

before
TTA CCG GCC ATC

after
TTA CCG GCC ATC **CAT**

CAT broke away from another chromosome.

Chromosome mutations

If a complete chromosome is added or deleted, this is a **chromosomal mutation**. Sometimes something goes wrong during meiosis and both members of a homologous pair of chromosomes move to the same pole. This produces a gamete with an extra chromosome and, after fertilisation, the zygote has an extra chromosome. This is called aneuploidy. A example is Down's syndrome where a person has an additional chromosome, totalling 47 in each nucleus rather than the usual 46.

If the spindle fails altogether, then an individual can be produced with whole extra sets of chromosomes. This is called polyploidy and is important in plant evolution.

What causes mutations?

All organisms tend to mutate randomly, so different sections of DNA can appear to alter by chance. The appearance of such a random mutation is usually very rare, typically one mutation in many thousands of individuals in a population. The rate can be increased by **mutagens** such as:

- **Ionising radiation** – including ultra violet light, X-rays and α(alpha), β(beta) and γ (gamma) rays and neutrons. These forms of radiation tend to dislodge the electrons of atoms and so disrupt the bonding of the DNA which may re-bond in different combinations.
- **Chemicals** – including asbestos, tobacco, nitrous oxide, mustard gas and many substances used in industrial processes such as vinyl chloride. Many pesticides are suspected mutagens. Dichlorvos, an insecticide, is a proven mutagen. Additionally, colchicine is a chemical derived from the Autumn crocus, *Colchicium*, which stimulates the development of extra sets of chromosomes.

Are mutations harmful or helpful?

An individual mutation may be either harmful or helpful. When tobacco is smoked, it can increase the rate of mutation in some somatic cells. The DNA disruption can result in the formation of a cell which divides uncontrollably and causes the disruption of normal body processes. This is **cancer**, and can be lethal. The presence of certain genes called **oncogenes** is thought to increase the rate of cell division and lead to cancer.

Chrysanthemum plants have a high rate of mutation. A chrysanthemum grower will often see a new colour flower on a plant, e.g. a plant with red flowers could develop a side shoot which has a different colour, such as bronze. Most modern chrysanthemums appeared in this way, production being by asexual techniques.

Some mutated human genes have, through evolution, been successful. Many successful mutations contributed to the size of the cerebrum which is proportionally greater in humans than in other primates.

5.2 Selection and speciation

After studying this section you should be able to:

- *understand the process of natural selection*
- *predict population changes in terms of selective pressures*
- *understand a range of isolating mechanisms and how a new species can be formed*
- *understand the difference between allopatric and sympatric speciation*
- *explain the difference between natural selection and artificial selection*

LEARNING SUMMARY

Natural selection

OCR   5.1.2

Throughout the biosphere, communities of organisms interact in a range of ecosystems. Darwin travelled across the world in his ship, the *Beagle*, observing organisms in their habitats. In 1858 Darwin published *On the Origin of Species*. In this book he gave his theory of **natural selection**.

The key features of this theory are that as organisms interact with their environment:

- individual organisms of populations are not identical, and can **vary in both genotypes and phenotypes**
- **some organisms survive** in their environment other organisms **die** before reproducing, effectively being **deleted from the gene pool**
- surviving organisms **go on to breed** and **pass on their genes** to their offspring
- this **increases the frequency of the advantageous genes** in the population.

> Learn this theory carefully then apply it to the scenarios given in your examination. Candidates often identify that some organisms die and others survive, but few go on to predict the inheritance of advantageous genes and the consequence to the species.

Consider these factors

- Adverse conditions in the environment could make a species extinct, but a range of genotypes increases the chances of the species surviving.
- Different genotypes may be suited to a changing environment, say, as a result of global warming.
- A variant of different genotype, previously low in numbers, may thrive in a changed environment and increase in numbers.
- Where organisms are well suited to their environment they have adaptations which give this advantage.
- If other organisms have been selected against, then more resources are available for survivors.
- Breeding usually produces many more offspring than the mere replacement of parents.
- Resources are limited so that competition for food, shelter and breeding areas takes place. Only the fittest survive!

What is selective pressure?

Selective pressure is the term given to a factor which has a direct effect on the numbers of individuals in a population of organisms, for example:

> 'It is late summer and the days without rainfall have caused the grassland to be parched. There is little food this year.'

> In this example, the fact that the numbers of herbivores decrease is *another* selective pressure. This time numbers of predators may decrease.

Here the **selective pressure** is a **lack of food** for the herbivores. Species which are **best adapted** to this habitat **compete** well for the limited resources and go on to survive. Within a species there is a further application of the selective pressure as weaker organisms perish and the strongest survive.

81

Mutations are random

New genes can appear in a species for the first time, due to a form of mutation. Over thousands of years, repeated natural selection takes place, resulting in superb adaptations to the environment.

- The Venus fly trap with its intricate leaf structures captures insects. The insects decompose, supplying minerals to the mineral deficient soil.
- Crown Imperial lilies (*Fritillaria*) produce colourful flowers, and a scent of stinking, decomposing flesh. Flies are attracted and help pollination.
- The bee orchid flower is so like a queen bee that a male will attempt mating.

Considering these examples, it is no wonder that candidates seem to consider that the organisms actively adapt to develop in these ways. They suggest that the organisms themselves have control to make active changes. **This is not so! There is no control, no active adaptation.**

> *New genes appear by CHANCE!*

KEY POINT

Selective pressures and populations

To find out more about the effects that selective pressures can have, the **normal distribution** must be considered. The distribution below is illustrated with an example.

The mean value is at the peak. There are fewer tall and short individuals in this example. A taller plant intercepts light better than a shorter one.

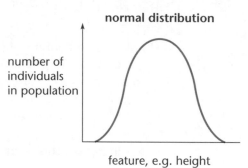

normal distribution

number of individuals in population

feature, e.g. height

The further distributions below show effects of selective pressures (shown by the blue arrows). Each is illustrated with an example.

Selective pressure at both ends of the distribution causes the extreme genotypes to die. This maintains the distribution around the mean value. Mean wing length is better for flight, better for prey capture.

Selective pressure results in death of slower animals. Many die out due to predators. Faster ones (with longer legs) pass on advantageous genes. Distribution moves to the right as the average individual is now faster.

Selective pressure results in the death of organisms around the mean value. In time this can lead to two distributions. Long fur is adapted to a cold temperature and short fur to a warm temperature. The mean is suited to neither extreme.

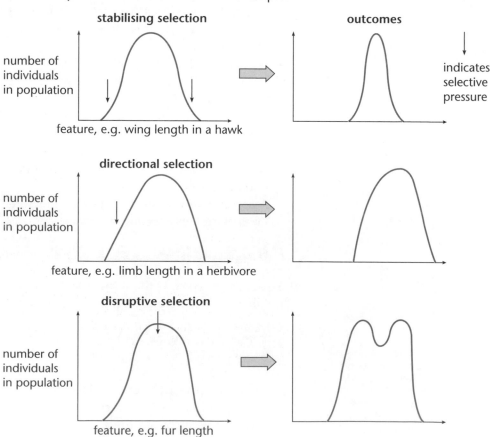

stabilising selection

number of individuals in population

feature, e.g. wing length in a hawk

outcomes

↓ indicates selective pressure

directional selection

number of individuals in population

feature, e.g. limb length in a herbivore

disruptive selection

number of individuals in population

feature, e.g. fur length

Speciation

OCR   5.1.2

The previous example of disruptive selection showed how two extreme genotypes can be selected. Continued selection against individuals around the former mean genotype finally results in two discrete distributions. This division into two groups may be followed by, for example, advantageous mutations. There is a probability that, in time, the two groups will become incompatible and unable to breed successfully. They have become a new species. The **development of new species is called speciation**.

To enable enough genetic differences to build up between the two groups, they must be isolated to stop them breeding. This can happen in a number of ways.

Geographical isolation

This will help you. Different finches evolved on different islands, but they did have a common ancestor.

This takes place when a population becomes divided as a result of a physical barrier appearing. For example, a land mass may become divided by a natural disaster like an earthquake or a rise in sea level. Geographical isolation followed by mutations can result in the formation of new species. This can be illustrated with the finches of the Galapagos islands. There are many different species in the Galapagos islands, ultimately from a common ancestral species. Clearly new species do form after many years of geographical isolation. This is **allopatric speciation**.

Reproductive isolation

A new pheromone is produced by several antelopes as a result of a mutation. The mainstream individuals refuse to mate as a result of this scent. An isolated few do mate. This is reproductive isolation.

This is a type of genetic isolation. Here the formation of a new species can take place in the same geographical area, e.g. mutation(s) may result in reproductive incompatibility. A new gene producing, for example, a hormone, may lead an animal to be rejected from the mainstream group, but breeding may be possible within its own group of variants. The production of a new species by this mechanism is known as **sympatric speciation**.

Artificial selection

OCR   5.1.2

In natural selection, the selection pressure comes from the organism's environment. In **artificial selection**, humans choose which organisms are allowed to reproduce. This is **selective** breeding to improve specific domesticated animals and crop plants.

Important points are:

Artificial selection is not the only way to improve animals and plants. Genetic modification is another method. A variety of soya bean plants now has resistance to selective herbicides.

- people are the **selective agents** and choose the parent organisms which will breed
- the organisms are chosen because they have **desired characteristics**
- the aim is to incorporate the desired characteristics from both organisms in their offspring
- the offspring must be **assessed** to find out if they have the desired combination of improvements (there is **no guarantee** that a cross will be successful!)
- offspring which have suitable improvements are used for breeding, the others are deleted from the gene pool (not allowed to breed).

Most modern crops have been produced by artificial selection. Modern wheat is one example. The Brussels sprout variety opposite was produced in this way. Many trials were carried out before the new variety was offered for sale.

Can you suggest four excellent features offered by this new variety?

Brilliant NEW FOR 2001
F1 Hybrid A brand new early cropping variety which produces dense, dark green buttons of excellent quality in September and October. Suitable for a wide range of soil types it also has a high resistance to powdery mildew and ring spot. Good for freezing.   2152 pkt £2.10

All modern racehorses have been artificially selected. Champion thoroughbred horses are selected for breeding on the basis of success in races. Only the best racehorses are actually entered in races. The fastest horses, at various distances, win races and the right to breed. Continual improvement results as the gene pool is consistently strengthened. Modern dairy cows have also been produced by artificial selection.

Sample question and model answer

The graphs below show the height of two pure breeding varieties of pea plant, Sutton First and Cava Late.

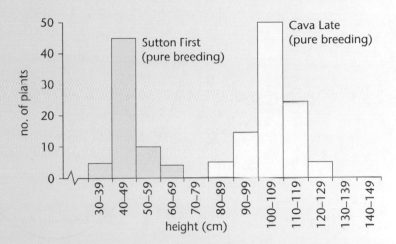

Continuous variation can confuse you sometimes when examiners display the data in categories as histograms. **This is not discontinuous!**

(a) (i) Which types of variation are shown by the pea variety, Sutton First? Give evidence from the bar graph to support your answer. [4]

Continuous variation – this is shown by the increase across the distribution (even though the peas are pure breeding).

Environmental variation – shown by the range of different heights.

(ii) Which type of variation is shown **between** varieties Sutton First and Cava Late? Give evidence from the bar graphs to support your answer. [2]

Discontinuous variation – the two distributions are separate and do not intersect.

(iii) Both Sutton First and Cava Late have compatible pollen for cross-breeding. Suggest why they do **not** cross breed. [1]

As implied by the names, Sutton First flowers before Cava Late, so that the flowers are not ready at the same time.

When you are asked to 'suggest', then a range of different plausible answers are usually acceptable.

(b) Plant geneticists considered that many years ago the two varieties of pea had the same ancestor.

(i) Suggest what, in the ancestor, resulted in the difference in height of the two varieties? [1]

mutation

(ii) Suggest what caused this change. [1]

radiation/random processes

(c) (i) Define polygenic inheritance. [1]

The inheritance of a feature controlled by a number of genes (not just a gene at one locus).

(ii) Which type of variation is a consequence of polygenic inheritance? [1]

continuous variation

Practice examination questions

1   The graph shows the birth weight of babies born in a London hospital between 1935 and 1946. It also shows the chance of the babies dying within two months of birth.

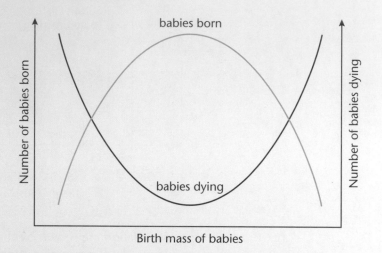

(a) What type of variation is shown by the birth weight of the babies? [1]

(b) What factors decide the birth weight of a baby? [2]

(c) Scientists argue that the information in the graphs shows that stabilising selection acts on the birth weight of babies. Explain why they think this. [3]

(d) The data was collected between 1935 and 1946. Modern medical techniques may have altered the selection pressure on the birth weight of babies. Explain why. [2]

[Total: 8]

2   (a) Explain the difference between allopatric and sympatric speciation. In each instance use an example to illustrate your answer. [6]

(b) How is it possible to find out if two female animals are from the same species? [2]

[Total: 8]

Chapter 6

Biotechnology and genes

The following topics are covered in this chapter:

- Growing microorganisms
- Mapping and manipulating genes
- Cloning

6.1 Growing microorganisms

After studying this section you should be able to:

- describe how microorganisms are grown in ideal conditions in fermenters
- describe the main features of batch and continuous culture

LEARNING SUMMARY

Microorganisms and fermenters

OCR ▶ 5.2.2

Before describing how substances are commercially produced it is necessary to consider the meaning of the term, **biotechnology**.

> Biotechnology is the use of organisms and biological processes to supply nutrients, other substances and services to meet human needs. **Fermentation** is a key process in biotechnology using microorganisms to produce traditional products such as ethanol and more recently substances such as pharmaceutical chemicals and the enzymes for biological washing powder.
>
> KEY POINT

Modern industrial fermenters

New transgenic organisms are continually being developed. Many hit the headlines in the media. Be aware that the examiners may use these high-profile organisms in questions. Do not become disorientated – the principle is always the same.

- There are several different types of fermenter used to grow microorganisms on a large scale. They all have the common purpose of producing food, or chemicals such as antibiotics, hormones, or enzymes. The fermenter below shows a typical design.

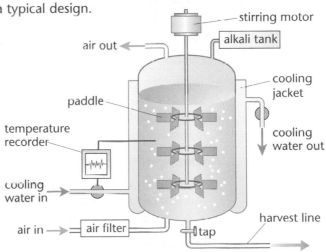

Conditions inside fermenters should be suitable for the optimal metabolism and rapid reproduction of the microorganisms. Products should be harvested without contamination. Note the conditions which need to be controlled.

- Fermenters are sterilised using steam before adding nutrients and the microorganisms used during the process. Conditions are **aseptic**.
- Nutrients which are specifically suited to the needs of the microorganisms are supplied.
- Air is supplied if the process is aerobic. This must be filtered to avoid

contamination from other microorganisms.

- Temperature must be regulated to keep the microorganisms' enzymes within a suitable range. An active 'cooling jacket' and heater, both controlled via a thermostat, enable this to be achieved.
- pH must remain close to the optimum. Often the development of low pH during fermentation would result in the process slowing down or stopping. The addition of alkaline substances allows the process to continue and maximises yield.
- Paddle wheel mixing or 'bubble agitation' make sure that the microorganisms meet the required concentrations of nutrients and oxygen.

This graph shows the rate of production of a primary metabolite. These are produced by a microbe as part of its normal growth. Secondary metabolites such as antibiotics are usually produced after the main growth phase.

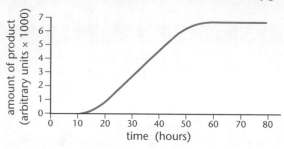

Batch culture

Batch culture takes place in a **closed** vessel. The microorganism is allowed to grow and then stopped and the product is removed.

Advantages

- If the culture becomes contaminated in any way, just one batch is spoiled.
- The fermenter can be used for a variety of fermentation processes, e.g. different antibiotics.

Disadvantages

- At the end of every production period, **shut down** takes place. The vessel needs to be cleaned and re-sterilised. This lost time can be expensive to the company.
- Often the product, waste substances and unused nutrients are mixed together, e.g. in penicillin production. Product removal is made more difficult by these contaminants.

Continuous culture

Continuous culture takes place in an open fermenter.

- The fermenter is steam sterilised.
- Regular amounts of sterile nutrients are added.
- At the same time regular amounts of product are removed.
- Optimum levels of pH, oxygen, nutrients and temperature are maintained.

Advantages

- The rate of growth of the microbial population is kept at a maximum level: this is known as the **exponential rate**.
- There is **no** regular pattern of **shut down**.

Disadvantages

- Maintaining the levels at **optimal levels** is difficult.
- **Regular** sampling is necessary for quality control, ensuring that chemicals are in equilibrium, and contaminants are absent.
- There is more chance of contaminants entering due to regular input and output.

Immobilised enzymes

After completion of an enzyme catalysed reaction the enzyme remains unchanged and can be used again. Unfortunately the enzyme can contaminate the product, as it can be difficult to separate out from the reaction mixture. For this reason

immobilised enzymes have been developed. They are used as follows:

- enzymes are attached to insoluble substances such as resins and alginates
- these substances usually form membranes or beads and the enzymes bind to the outside
- substrate molecules readily bind with the active sites and the normal reactions go ahead
- the immobilised enzymes are easy to recover, remaining in the membranes or beads
- there is no contamination of the product by free enzymes
- expensive enzymes are re-used
- processes can be continuous unlike batch, where the process is stopped for 'harvesting'.

Progress check

State **two** differences between continuous and batch production.

- Batch production is shut down on a more regular basis.
- In continuous production, nutrients are added on a regular basis, and products are removed in similar quantities. In batch production product retrieval is at the end rather than during the process.

6.2 Cloning

After studying this section you should be able to:

- *describe how plants can be cloned*
- *explain why it is harder to clone animals*
- *describe the possible uses of cloning*

LEARNING SUMMARY

Producing clones

OCR   5.2.1

As discussed on page 63 the cells in the early embryo are called **totipotent**. This means that they can develop into any type of cell. However, as the embryo develops, cells become specialised and lose this ability. In mature plants, many cells remain totipotent. This means that it is quite easy to produce new plants from sections of plant tissue. The plants that are produced are called clones. In mature animals these totipotent or **stem cells** are harder to find.

> Clones are genetically identical to their parent. Plants can clone themselves naturally or can be cloned artificially by processes such as **tissue culture** or **micropropagation**.
>
> **KEY POINT**

Advantages of micropropagation

- Generating new plants from the apical meristem tissue eliminates many plant viruses, so usually, virus-free plants are produced.
- If the material is available the process can take place at any time of the year.
- Even a tiny explant or callus can be cut into pieces and sub-cultured.

The diagrams below outline the process.

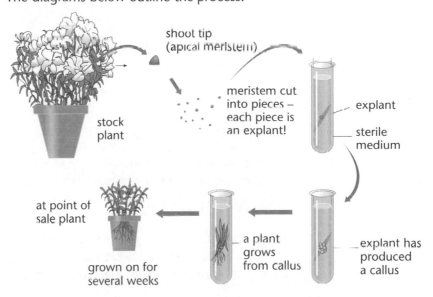

shoot tip (apical meristem)

stock plant

meristem cut into pieces – each piece is an explant!

explant

sterile medium

explant has produced a callus

a plant grows from callus

grown on for several weeks

at point of sale plant

89

Cloning animals

Cloning animals is much more difficult because of the problem of finding suitable stem cells. There are two possibilities:

- Take an early embryo and split up the ball of cells. At this early stage the cells will be able to develop into separate cloned individuals.
- It is now possible to clone many animals from adult body cells. This is performed by nuclear transfer and was first used to produce Dolly the sheep in 1996. The diagram shows how it is done.

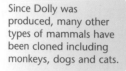

Since Dolly was produced, many other types of mammals have been cloned including monkeys, dogs and cats.

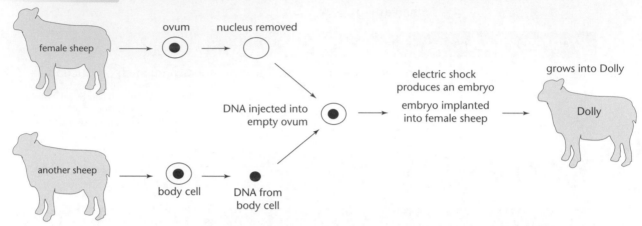

Uses of cloning

Cloning animals by splitting up an embryo may have limited uses. As the embryo has been produced by sexual reproduction, the genetic make-up of the clones is uncertain. Cloning animals from adult body cells may be much more useful. There are two main possibilities:

- **Reproductive cloning** This could be used to produce identical copies of endangered animals, animals with desired characteristics or even embryos for infertile human couples.
- **Therapeutic cloning** This may be used to provide a source of stem cells from the early embryo. The stem cells may be used to treat degenerative diseases. The embryo is then destroyed.

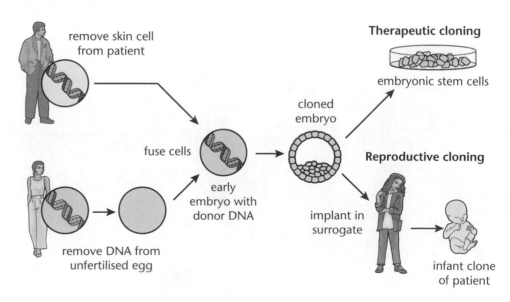

6.3 Mapping and manipulating genes

After studying this section you should be able to:

- *describe a range of techniques used in mapping genes*
- *describe how genes can be transferred between a range of different organisms*

Identifying genes

OCR   5.2.3

In order to identify the genes of an individual, a number of different processes are used.

Polymerase chain reaction

The polymerase chain reaction (PCR) is used to make numerous copies of a section of DNA. This is called **amplifying** the DNA. It uses the principle of semiconservative replication of DNA to produce new molecules that can in turn act as templates to produce more molecules. This therefore sets up a chain reaction. The enzyme DNA polymerase is used to copy the DNA.

Electrophoresis

This is used to separate sections of DNA according to their size.

Enzymes called restriction endonucleases can be used to cut up an organism's DNA.

- DNA sections are put into a well in a slab of agar gel.
- The gel and DNA are covered with buffer solution which conducts electricity.
- Electrodes apply an electrical field.
- Phosphate groups on DNA are negatively charged causing DNA to move towards the anode.
- Smaller pieces of DNA move more quickly down the agar track; larger ones move more slowly, leading to the formation of bands.

Genetic fingerprinting

PCR and electrophoresis have many applications. DNA is highly specific so the bands produced using this process can help with identification. In some crimes, DNA is left at the scene. Blood and semen both contain DNA specific to an individual. DNA evidence can be checked against samples from suspects. This is known as **genetic fingerprinting**. Genetic fingerprinting can be used in paternity disputes. Each band of the DNA of the child must correspond with a band from *either* the father or the mother.

Isolating genes

Along chromosomes are large numbers of genes. Scientists may need to identify and isolate a useful gene; one way of doing this is to use the enzyme **reverse transcriptase**. This is produced by viruses known as **retroviruses**. Reverse transcriptase has the ability to help make DNA from mRNA.

Stage 1

When a polypeptide is about to be made at a **mRNA** ribosome, reverse transcriptase allows a strand of its coding DNA to be made.

Stage 2

The single stranded DNA is parted from the mRNA.

Stage 3

The other strand of DNA is assembled using DNA polymerase.

Using this principle, the exact piece of DNA which codes for the production of a vital protein can be made.

Progress check

1   A length of DNA was prepared and then electrophoresis was used to separate the sections. The statements below describe the process of electrophoresis but they are in the wrong order. Write the letters in the correct sequence.

   A   electrodes apply an electrical field
   B   DNA sections are put into a well in a slab of agar gel
   C   smaller pieces of DNA move more quickly down the agar track with larger ones further behind
   D   the gel and DNA are then covered with buffer solution which conducts electricity
   E   restriction endonucleases can be used to cut up the DNA

2   Reverse transcriptase is an enzyme which enables the production of DNA from RNA. Work out the sequence of organic bases along the DNA of the following RNA sequence.

   A A U GCCCGGAUU

   2   RNA AAUGCCCGGAUU   DNA₂ AATGCCCGGATT   DNA₁ TTACGGGCCTAA
   1   E B D A C

The Human Genome Project

The Human Genome Project is an analysis of the complete human genetic make-up, which has mapped the organic base sequences of the nucleotides along our DNA.

A brief history

- 1977 Sanger devised DNA base sequencing.
- 1986 The Human Genome Project was initiated in the USA and the UK.
- 1996 30 000 genes were mapped.
- 1999 one billion bases were mapped including all of chromosome 22.
- 2000 chromosome 21 was mapped with the human genome almost complete.
- 2001 human genome mapping complete.

Some important points

- The genome project will sequence the complete set of over 100 000 genes.
- Only around 5% of the base pairs along the DNA actually result in the expression of characteristics. These DNA sequences are known as **exons**.
- 95% of DNA base sequences are not transcribed and do not appear to be involved in the expression of characteristics. These are known as **introns**.
- Introns do not outwardly seem to be responsible for characteristics. It is likely that they may be regulatory, perhaps in multiple gene role.

Effects of single nucleotide polymorphism

Example

5 base sequences from five people →

GTATAGCCGCAT	1
GTATAGCCGCAT	1
GTATAGCCGCAT	1
GTATAGCCGCCT	2
GTATAGCCGCCT	2

Version 1 = ●
Version 2 = ●

Proportion of the SNP in healthy members of population:

Proportion of the SNP in diseased members of population:

A greater incidence of an SNP in people with a disease may point to a cause.

Single nucleotide polymorphisms (SNPs)

Around 99.9% of human DNA is the same in all individuals. Merely 0.1% is different! The different sequences in individuals can be the result of **single nucleotide polymorphism**. One base difference from one individual to another at a site may have no difference. Up to a maximum of six different codons can code for one amino acid. An SNP will not necessarily have any effect.

Some SNPs do change a protein significantly. Such changes may result in genetic disease, resistance or susceptibility to disease.

How can the mapping of SNPs be useful?

- The mapping of SNPs along chromosomes signpost where base differences exist.
- Across the gene pool a pattern of SNP positions will be evident.
- There may be a high frequency of common SNPs found in the DNA of people with a specific disease.
- This highlights interesting sites for future research and will help to find answers to genetic problems.

Benefits obtained from the Human Genome Project

Ultimately, the human genome data will be instrumental in the development of drugs to treat genetic disease. Additionally, by analysis of parental DNA, it will be possible to give the probability of the development of a specific disease or susceptibility to it, in offspring. Fetal DNA, obtained through amniocentesis or by chorionic villi sampling, will give genetic information about an individual child.

Genetic counsellors will have more information about an individual than ever before. Companies will be able to produce 'designer drugs' to alleviate the problems which originate in our DNA molecules. Soon the race will begin to produce the first crop of drugs to treat or even cure serious genetic diseases. Look to the media for progress updates.

Manipulating DNA

OCR 5.2.3

Scientists have developed methods of manipulating DNA. It can be transferred from one organism to another. Organisms which receive the DNA then have the ability to produce a new protein. This is one example of **genetic engineering**.

> The genetic code is universal. This means that it is possible to move genes from one organism to another and the recipient organism may be from a different species. The DNA will still code for the same protein.

KEY POINT

Genes have now been transferred to and from many different types of organisms.

Here are some examples:

- From humans to bacteria: this technique produced the first commercially available genetically engineered product, insulin.
- From plants to plants: this technique has been used to produce GM crops such as Golden Rice that contain vitamin A.
- Into humans: this technique may be successful in treating genetic conditions such as cystic fibrosis (but it is not a cure).

Gene transfer to bacteria

The gene which produces human insulin was transferred from a human cell to a bacterium. The new microorganism is known as a **transgenic bacterium**. The process which follows shows how a human gene can be inserted into a bacterium.

The human gene for insulin is produced using an enzyme called **reverse transcriptase**. This converts the mRNA coding for insulin back into DNA. In this way all the introns are removed. Then the DNA can be inserted.

1 An enzyme known as **restriction endonuclease** cuts the DNA and the gene was removed. Each time a cut was made the two ends produced were known as 'sticky ends'.

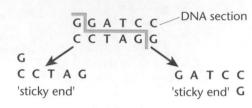

Restriction endonucleases are produced by some bacteria as a defence mechanism. They cut up the DNA of invading viruses. This can be exploited during gene transfer.

2 Circles of DNA called **plasmids** are found in bacteria.

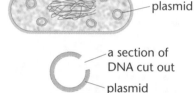

Note that **both** the donor DNA and recipient plasmid DNA are cut with the same enzyme. This allows the new gene to be a matching fit.

3 A plasmid was taken from a bacterium and cut with the same restriction endonuclease.

4 The human gene was inserted into the plasmid. It was made to fix into the open plasmid by another enzyme known as **ligase**.

Many exam candidates fail to state that the plasmids are cloned inside the bacterium.

5 The plasmid **replicated** inside the bacterium.

6 Large numbers of the new bacteria were produced. Each was able to secrete perfect human insulin, helping diabetics all over the world.

The bacteria themselves are also cloned. There may be two marks in a question for each cloning point!

Gene transfer to plants

Inserting genes into crop plants is becoming increasingly important in meeting the needs of a rising world population. One example of this is the production of a type of rice called Golden Rice. This contains a gene that produces vitamin A. The aim is to prevent vitamin A deficiency which can lead to blindness.

In **plants** there is an important technique which uses a **vector** to insert a novel gene. The vector is the bacterium *Agrobacterium tumefaciens*.

> *Agrobacterium tumefaciens*
> - This is a **pathogenic bacterium** which **invades** plants forming a gall (abnormal growth).
> - The bacterium contains **plasmids** (circles of DNA) which carry a gene that stimulates tumour formation in the plants it attacks.
> - The part of the plasmid which does this is known as the **T-DNA region** and can insert into any of the chromosomes of a host plant cell.
> - Part of the T-DNA controls the production of two growth hormones, auxin and cytokinin.
> - The extra quantities of these hormones stimulate rapid cell division, the cause of the tumour.
>
> **KEY POINT**

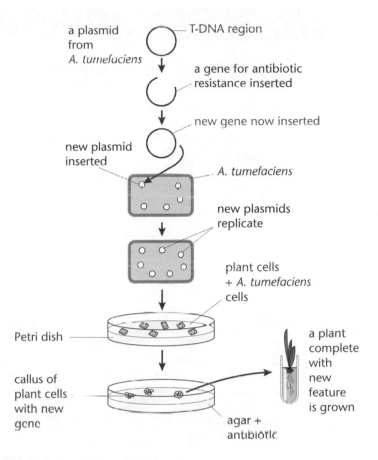

How can *Agrobacterium tumefaciens* be used in gene transfer?

The principle of using *A. tumefaciens* can be used in gene transfer to many different plants. Applications are at an early stage of development.

- Firstly, the DNA section controlling auxin and cytokinin was deleted, tumours were not formed, and cells of the plant retained their normal characteristics.
- A gene which gave the bacterial cell **resistance to a specific antibiotic** was inserted into the T-DNA position.
- The **useful gene** (e.g. the gene for vitamin A) was **inserted into a plasmid.**
- Plant cells, minus cell walls, were removed and put into a Petri dish with nutrients and *A. tumefaciens*, which contained the engineered plasmids.
- The cells were **incubated** for several days, then transferred to another Petri dish containing nutrients plus the specific antibiotic.
- **Only plant cells with antibiotic resistance and the desired gene grew.**
- Any surviving cells grew into a callus, from which an adult plant formed, complete with the transferred gene.

Inserting genes into humans

Genes can also be inserted into animals to prevent their organs being rejected if used for transplants into people. This is called **xenotransplantation.**

The idea of changing a person's genes in order to cure genetic disease is called **gene therapy.**

There are two main possibilities:

- **Somatic cell therapy** in which the genes are inserted into the cells of the adult where they are needed.
- **Germ line gene therapy** involves changing the genes of the gametes or early embryo. This means that all the cells of the organism will contain the new gene.

Sample question and model answer

The diagrams below show the transfer of a useful gene from a donor plant cell to the production of a transgenic crop plant. The numbers on the diagram show the stages in the process.

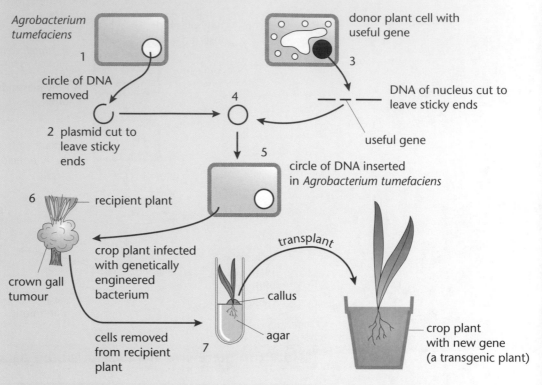

Look out for transgenic stories in the media. The principles are often the same. This will prepare you for potentially new ideas in your 'live' examinations. You could encounter the same account!

(a) Give the correct name for the circle of DNA found in the bacterium, *A. tumefaciens*. [1]

plasmid

(b) The same enzyme was used to cut the DNA of the bacterium and of the plant cell.

(i) Name the type of enzyme used to cut the DNA. [1]

restriction endonuclease

(ii) Explain why it is important to use exactly the same enzyme at this stage. [2]

The same enzyme produces the same sticky ends.

Complementary sticky ends on the donor gene bind with the sticky ends of the plasmid.

This question covers key techniques in gene transfer. Be prepared for your examination.

(iii) Which type of enzyme would be used to splice the new gene into the circle of DNA? [1]

ligase

(c) How was the new gene incorporated into the DNA of the crop plant cells? [2]

Crop plant infected by genetically engineered bacterium.

The DNA of bacterium causes a change in the DNA of the crop plant to produce the gall or tumour cells.

(d) How would you know if the gene had been transferred successfully? [1]

The feature would be expressed in the transgenic plants.

Practice examination questions

1 The diagram below shows an industrial fermenter used to produce the antibiotic, penicillin.

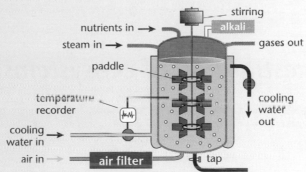

(a) Describe **three** ways in which aseptic conditions are achieved in the fermenter. [3]

(b) If the air filter failed, explain what would be the likely effect inside the fermenter. [3]

[Total: 6]

2 A new genetically modified soya bean plant has been developed. It has a new gene which prevents it from being killed by herbicide (weed killer).

(a) Describe the stages which enable a gene to be transferred from one organism to another. [5]

(b) Suggest how the genetically modified soya plants could result in higher bean yields. [3]

[Total: 8]

3 The graph below shows the level of product secreted by microorganisms in a commercial fermenter.

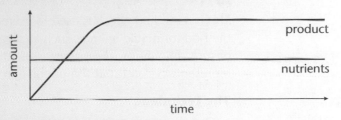

(a) Account for the shape of the graph. [1]

(b) Which type of culture, batch or continuous, took place in this fermenter?
Give **two** reasons for your choice. [2]

[Total: 3]

Chapter 7

Ecology and populations

The following topics are covered in this chapter:

- Investigation of ecosystems
- Behaviour
- Colonisation and succession

7.1 Investigation of ecosystems

After studying this section you should be able to:

- use the capture, mark, recapture technique to assess animal populations
- use quadrats to map the distribution of organisms
- understand the factors that affect the distribution of organisms
- describe conservation techniques and methods of population control

LEARNING SUMMARY

Measurement in an ecosystem

OCR ▶ 5.3.1-2

The study of ecology investigates the inter-relationships between organisms in an area and their environment. The area in which organisms live is called a **habitat**. The combination of the organisms that live in a habitat and the physical aspects of the habitat is called an **ecosystem**.

Estimating populations

All the individuals of one species living together in a habitat are called a **population**. The size of plant populations can be estimated by using a quadrat placed at random. Animals, however, do not tend to stay still for long enough to be sampled using a quadrat. The population size of an animal species can be estimated by using capture–recapture.

Capture, mark, release, recapture

This is a method which is used to estimate animal populations. It is an appropriate method for motile animals such as shrews or woodlice. The ecologist must always ensure minimum disturbance of the organism if results are to be truly representative and that the population will behave as normal.

The technique

- Organisms are captured, *unharmed*, using a quantitative technique.
- They are counted then discretely marked in some way, e.g. a shrew can be tagged, a woodlouse can be painted (*with non-toxic paint*).
- They are released.
- Organisms from the same population are recaptured, and another count is made, to determine the number of marked animals and the number unmarked.

Before using the technique you must be assured that:

- there is no significant migration
- there are no significant births or deaths
- marking does not have an adverse effect, e.g. the marking paint should not allow predators to see prey more easily (or vice versa)
- organisms integrate back into the population after capture.

Remember that the method is suitable for large population size only.

The calculation

S = total number of individuals in the total population.

S_1 = number captured in sample one, marked and released, e.g. 8.

S_2 = total number captured in sample two, e.g. 10.

S_3 = total marked individuals captured in sample two, e.g. 2.

$$\frac{S}{S_1} = \frac{S_2}{S_3} \quad \text{so, } S = \frac{S_1 \times S_2}{S_3} \quad \text{population} = \frac{8 \times 10}{2} = 40 \text{ individuals}$$

Remember the equation carefully. You will **not** be supplied with it in the examination, but you will be given data.

Measuring the distribution of organisms

This can be measured using another quadrat technique called a belt transect. This method should be used when there is a **transition** across an area, e.g. across a pond or from high to low tide on the sea shore. Use belt transects where there is **change**. The belt transect is a line of quadrats. In each quadrat a measurement such as density can be made. One transect is not enough! Always do a number of transects then find an average for quadrats in a similar zone.

A bar graph would be used to show the **distribution** of plant species across the pond. Note that there would be more than just two species. The graphs show how you could illustrate the data. Clearly flag irises occupy a different niche to water lilies.

A simplified results table

Quadrat no.	flag Iris	water lily
1	10	0
2	7	0
3	1	0
4	0	5
5	0	4
6	0	0
7	0	5
8	0	3
9	0	0
10	1	0
11	8	0
12	4	0

This is just one belt transect. A number would be used and an average taken for each corresponding quadrat.

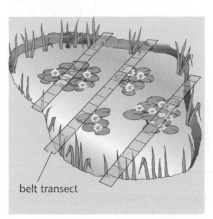

belt transect

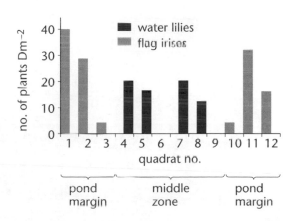

Other uses of quadrats

Quadrats can also be used to survey animal populations. It is made easier if the organisms are **sessile** (*they do not move from place to place*), e.g. barnacles on a rock. In a pond the belt transect could be coupled with a kick sampling technique. Here rocks may be disturbed and escaping animals noted. Adding a further technique can help, such as using a catch net in the quadrat positions. The principle here is that the techniques are **quantitative**.

KEY POINT

Factors that determine population size

Graphical data can show relative numbers and distribution of organisms in a habitat. The ecologist is interested in the factors that determine the size and distribution of organisms.

These factors can be **biotic** or **abiotic**.

Abiotic factors are non-living factors. They include:

- carbon dioxide level
- oxygen level
- pH
- light intensity
- mineral ion concentration
- level of organic material.

Biotic factors are living factors.

They include:

- **Competition** This occurs when organisms are trying to get the same resources. There are two types. **Interspecific competition** takes place when **different** species share the same resources. **Intraspecific competition** takes place when the **same** species share the same resources.
- **Predation** This involves feeding relationships.

Predators and prey

There can be many examples of this type of relationship in an ecosystem. **Primary consumers** rely on the **producers**, so a flush of new vegetation may give a corresponding increase in the numbers of primary consumers. Predators which eat the primary consumers may also follow with a population increase. Each population of the ecosystem may have a **sequential effect** on other populations. Ultimately, the ecosystem is in **dynamic equilibrium** and has limits as to how many of each population can survive, i.e. its **carrying capacity**.

Note that graphs are often given in predator–prey questions. A flush of spring growth is often responsible for the increase in prey. Plant biomass may not be shown on the graph! Candidates are expected to suggest this for a mark. Also remember that as prey increase, their numbers will go down when eaten by the predator. Predator numbers rise after this.

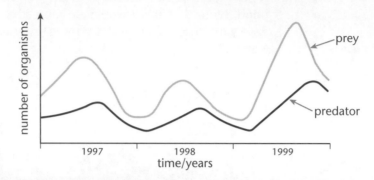

Progress check

(a) List the abiotic factors you may need to measure in a pond survey.

(b) How could you take measurements most efficiently, over a 24-hour period?

(b) Use of environmental probes, interface and computer.

(a) Oxygen, carbon dioxide, pH, light, temperature, mineral ions.

Ecological conservation

OCR S.3.2

In a world where human population increase is responsible for the destruction of so many habitats, it is necessary to retain as many habitats as possible. Ecological surveys report to governments and difficult decisions are made. Fragile habitats like the bamboo woodlands of China support a variety of wildlife. Conservation areas need to be kept and maintained to prevent extinction of organisms at risk. In the UK we have **sites of special scientific interest** (SSSI) which are given government protection.

Conservation requires management

Although the word 'conservation' implies to 'keep' something as it is, much effort is needed. An area of climax vegetation, e.g. oak woodland, is less of a problem, since it will not change if merely left to its own devices. However, many of the seral stages, e.g. birch woodland along the route to climax, require much maintenance.

Whilst **conservation** usually involves management in order to maintain biodiversity, **preservation** involves protecting areas in their untouched state.

Animal populations need our help, especially when it is often by our own introduction that specific species have colonised an area. Deer introduced into a forest may thrive initially but due to an efficient reproductive rate exceed the carrying capacity of the habitat. **Carrying capacity** is the population of the species which can be adequately supported by the area.

Sometimes herbivores could destroy their habitat by overgrazing, and so must be **culled**. **Predators** could be introduced to reduce numbers, but they also may need culling at some stage. **Difficult decisions** need to be taken. In the aquatic habitats similar problems exist. Cod populations in the North Sea are being reduced by over-fishing. Agreements have been made by the EU to **reduce fishing quotas** and create **exclusion zones** to **allow fish stocks to recover**. Even before this agreement, smaller fish had to be returned to the sea after being caught to increase the chances of them growing to maturity and breeding successfully.

Endangered species require protection

All over the world many animals and plants are at the limits of their survival. The World Wide Fund for Nature is a charity organisation which helps. The organisation receives support from the public and artists such as David Shepherd. He gives donations from the sale of all of his wildlife paintings, helping to maintain the profile of animals so that we invest in survival projects like protected reserves.

7.2 Colonisation and succession

After studying this section you should be able to:

- *understand how colonisation is followed by changes*
- *understand how colonisation and succession lead to a climax community*

How colonisation and succession take place

OCR 5.3.1

Any area which has never been inhabited by organisms may be available for **primary succession**. Such areas could be a garden pond filled with tap water, lava having erupted from a volcano, or even a concrete tile on a roof. The latter may become colonised by lichens. Occasionally an ecosystem may be destroyed, e.g. fire destroying a woodland. This allows **secondary succession** to begin, and signals the reintroduction of plant and animal species to the area.

> Colonisation and succession also take place in water. Even an artificial garden pond would be colonised by organisms naturally. Aquatic algae would arrive on birds' feet.

- **Pioneer species (primary colonisers)** begin to exploit a 'new' habitat. Mosses may successfully grow on newly exposed heathland soil. These are the **primary colonisers** which have adaptations to this environment. Fast germination of spores and the ability to grow in waterlogged and acid conditions aid rapid colonisation. These plants may support a specific food web. In time, as organic matter drops from these herbaceous colonisers it is decomposed, nutrients are added to the soil and acidity increases. In time, the changes caused by the primary colonisers make the habitat unsuitable.

- Conditions unsuitable for primary colonisers may be ideal for other organisms. In heathland, mosses are replaced by heathers which can thrive in acid and xerophytic (desiccating) conditions. This is **succession**, where one community of organisms is replaced by another. In this example, the secondary colonisers have replaced the primary colonisers; this is known as **seral stage 1** in the succession process. Again, a different food web is supported by the secondary colonisers.

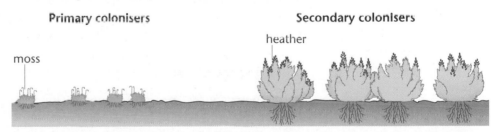

Primary colonisers   Secondary colonisers

moss   heather

- At every seral stage, there are changes in the environment. The **second seral stage** takes place as the tertiary colonisers replace the previous organisms. In heathland, the new conditions would favour shrubs such as gorse and bilberry plus associated animals.

- The shrubs are replaced in time with birch woodland, the **third seral stage**. Eventually, acidic build up leads to the destruction of the dominant plant species.
- Finally, conditions become suitable for a dominant plant species, the oak. Tree saplings quickly become established. Beneath the oak trees, grasses, ferns, holly and bluebells can grow as a balanced community. This final stage is **stable** and can continue for hundreds of years. This is the **climax community**. Associated animals survive and thrive alongside these plant resources. Insects such as gall wasps exploit the oak and dormice eat the wasp larvae.

Jays are birds which eat some acorns but spread others which they store and forget. The acorns germinate; the woodland spreads.

Climax community

oak woodland

In Britain, an excellent example of a climax community is Sherwood Forest where the 'Major Oak' has stood for 1000 years. Agricultural areas grow crops efficiently by **deflecting succession**. Plants and animals in their natural habitat are 'more than a match' for domesticated crops. Herbicides and pesticides are used to stop the invaders!

7.3 Behaviour

After studying this section you should be able to:

- *describe innate behaviour, kinesis and taxis*
- *understand habituation and imprinting*
- *describe a range of territorial behaviour*

LEARNING SUMMARY

The behaviour of organisms

OCR 5.4.3

Organisms respond to the biotic and abiotic factors of their environment. Biotic factors include response to other species, e.g. the feeding behaviour of grouse from heather on moorlands and the use of the heather to hide from predators. The grouse also respond to each other, e.g. in courtship display. There are different types of behaviour but they are either innate or learned.

Innate behaviour

This behaviour is '**pre-programmed**' by an organism's **genes**. When analysing behaviour it is difficult to determine whether it is innate or learned.

It is safe to say that immediately after the birth of a baby the 'sucking' action to obtain milk from the mother's mammary glands is innate. Similarly, the pecking behaviour of a chicken, while still in an egg, to break the shell, must be innate. As an animal gets older it may well develop patterns of behaviour learned from its experiences. It becomes more and more **difficult to categorise** the behaviour.

Kinesis

This takes place when the response of an organism is **proportional to the intensity of a stimulus**. Kinesis takes the form of an **increase in movement**, but this is **non-directional**. An example of kinesis is shown by woodlice. Intense heat which would harm the woodlice causes them to increase speed and move in random directions. In this way some of the population have a **greater chance of survival** by finding shelter.

Woodlice also respond to a dry environment by increasing random movements but slow down if they reach high humidity.

Taxis

This is a **directional response to a stimulus**. It can be a **positive taxis**, towards, or **negative taxis**, away. An example can be seen using a microscope to observe a group of living specimens of *Euglena viridis*. This is a protoctistan which photosynthesises. Individuals swim to an air bubble and cluster around to obtain maximum CO_2 for photosynthesis. This is **positive chemotaxis** because the organism moves towards the CO_2 source.

Learning

This takes place when an organism changes behaviour as a result of experience within the environment. As a result of the experience, future behaviour becomes modified. For example, a pupil misbehaves and is placed in detention. The pupil learns (hopefully!) that the behaviour should not be repeated. The detention is negative reinforcement. Perhaps positive reinforcement is better to support good behaviour.

Conditioned reflexes

Pavlov experimented with dogs.

Pavlov's dogs learnt to associate one stimulus to another. This is now described as **classical conditioning**.

If an animal learns to associate an action with getting a reward or punishment then this is called **operant conditioning**.

- He checked that the group of dogs did not produce saliva when he rang a bell at a time not related to feeding. (**Control**)
- He fed groups of dogs at a specific time each day.
- He measured the amount of saliva produced just before they were fed.
- He then began to ring a bell just before giving the food.
- The dogs began to salivate profusely.
- The bell would elicit exactly the same response as the original stimulus (the food).
- After a while the level of salivation decreased if the food reward was not given.
- Without **positive reinforcement** the level of response would finally disappear completely.

We are conditioned to respond to advertising in a similar way. A cola drink advertisement uses the latest popular song and glamorous models. We go to the supermarket and respond by buying the product, relating it to the pleasurable experience of the advertisement. Repeat purchases will only continue if the taste of the cola elicits a positive taste perception.

Habituation

This takes place when an organism is subjected to a **stimulus which is not harmful or rewarding**. As a result of continued subjection to a stimulus a **response will gradually decrease** and can finally disappear completely. A farmer puts an electronic bird scarer into a field. Birds are frightened off by frequent 'bangs'. They return, gradually getting closer and finally learn that the scarer is non-threatening. Soon they feed close to the scarer which has no effect. This is **habituation**.

Advertisements have a short 'shelf-life'. Continued exposure to the same advertisement results in habituation so that the response decreases. No wonder media advertising is replaced every few weeks!

The Sand Hill crane and imprinting

This endangered species is reared in incubators and re-introduced into the wild. There is a problem, though. Young cranes would imprint upon humans, so when re-introduced into the wild, it would move towards people. This would be dangerous, so each day, keepers dress up in 'crane' uniforms. In the wild the birds then move towards groups of adult cranes.

Imprinting

This takes place during the very early life of an organism, e.g. a chick emerges from its egg shell and immediately **bonds** with an object close by. In nature, this will normally be the mother. The mother will impart useful behavioural patterns to the chick, thus having **survival value**. From an incubator, the focus of the imprinting would be a human. The imprinting behaviour is that the chick, in this instance, will follow the human or any object to which it is first exposed.

Latent learning

If an animal encounters a new habitat it will investigate the area, learning to find its way around. This information is not of immediate use but may become useful in the future if for example, it is surprised by a predator. This type of learned behaviour is called latent learning.

Insight learning

Insight learning is the most advanced form of learning and it involves animals being able to predict the outcome of their actions. One of the most famous sets of observations was made on chimpanzees. The chimps could not reach some bananas that were outside their cage, even when provided with sticks. The chimps then learned to join the sticks together to enable them to reach the bananas.

Territorial behaviour

Populations of organisms living in an area can benefit from territorial behaviour. Too many animals of the **same species**, living in an area, **competing** for the same food would put the whole population in danger. Many species display territorial behaviour which prevents this outcome.

Examples are given to illustrate **principles**. It is unlikely that you will be given the same examples in your examination. Apply the principles to the given data.

Defence of the territory

- Animals are often **aggressive to members of the same species**, outside of the same family group.
- Territory is demarcated in a variety of ways, such as marking with **urine**, **faeces** or **scent**. Birds use **song**, whereas other animals have characteristic **calls**. Excluding others in these ways **can prevent physical confrontation** which often results in injury.
- It is an advantage to the species to have a **feeding range** which excludes others. This increases the chances of there being **enough food for the family group**.
- The apportioning of territories serves as **density dependent regulation** so that the best use is made of existing resources.
- A further advantage is that a territory marks out a designated **mating area**. Other males will usually remain outside the zone. Offspring have protection for their early development.

What is the advantage of male aggression to other males in a population?

The fittest organisms need to pass on their advantageous genes to offspring. In deer herds, a dominant stag (male) is challenged by a younger male occasionally. Antler to antler fights take place and there is potential damage to both stags. In time a new dominant stag takes over the family group and now has exclusive mating rights with a group of hinds (females). This behaviour ensures the male reproductive role involves only the strongest males. The gene pool is improved!

Courtship behaviour

This behaviour is species dependent and courtship display is anchored in the genes. Courtship rituals are very important to ensure that:

- the opposite sexes recognise each other
- the animals will mate with organisms from the same species (mating will produce fertile offspring)
- the act of mating is synchronised with the oestrous cycle. In pigs, a boar is always ready to mate but a sow is only receptive to him at ovulation. She produces pheromone attractants to encourage the boar.

Sample question and model answer

Read the passage, then answer the questions below.

line 1   Around the UK coast there are two species of barnacle, *Chthamalus stellatus* and *Balanus balanoides*. Both species are sessile, living on rocky sea shores.

The adult barnacles do not move from place to place but do reproduce
line 5   sexually. They use external fertilisation. Larvae resemble tiny crabs and are able to swim. At a later stage these larvae come to rest on a rock where they become fixed for the remainder of their lives.

The barnacles are only able to feed while submerged.

Adult *Chthamalus* are found higher on the rocks than *Balanus* in the adult
line 10   form as shown in the diagram below. Scientists have shown that the larvae of each species are found at all levels.

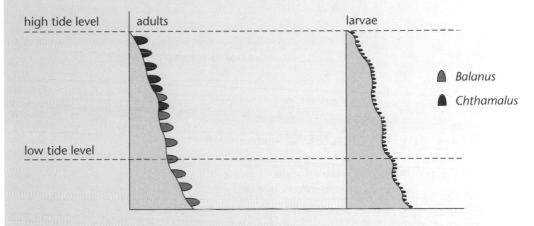

(a) Name the genus for each barnacle.   [1]

   Chthamalus and Balanus

(b) What does the term sessile mean? (line 2)   [1]

   is not motile, i.e. does not move from place to place

(c) Suggest how it is possible for neighbouring *Balanus* individuals to breed sexually (line 5) with each other even though they are sessile.   [1]

   produce sperms which swim through the water

(d) Explain **one** advantage of the larvae being motile.   [2]

   able to colonise new areas

   where there may be more nutrients / food

(e) Which type of competition exists between *Chthamalus* and *Balanus*?   [1]

   interspecific competition

(f) Suggest an explanation for the distribution of each species of barnacle.   [6]

   The larvae are found at all tide levels; at a lower tide level the barnacles are submerged for longer; Balanus may grow at a faster rate; can compete for food better; Chthamalus, which die out at lower levels near higher tide level the barnacles are exposed to open air for longer; Balanus may not be adapted to withstand desiccation; whereas Chthamalus can withstand drying out; and so survive without the competition from Balanus.

Practice examination questions

1 Ecologists wished to estimate the population of a species of small mammal in a nature reserve.

- They placed humane traps throughout the reserve and made their first trapping on day one, capturing 16 shrews.
- They were tagged then released.
- After day four a second trapping was carried out, capturing 12 shrews.
- Five of these shrews were seen to be tagged.

(a) The ecologists must be satisfied of a number of factors before using the 'capture, mark, release, recapture' method. List three of these factors. [3]

(b) Use the data to estimate the shrew population.
Show your working. [2]

(c) Comment on the *level* of reliability of your answer. [1]

[Total: 6]

2 Complete the table below by putting a tick in an appropriate box. You may tick one or more boxes for each example.

	Type of behaviour			
	simple reflex	kinesis	positive taxis	negative taxis
A bolus of food reaches the top of our oesophagus and is swallowed.				
Insects move from a cold, dry area to a warm, humid one.				
Springtails (insects) are subjected to increasingly hot conditions, and react by increasing speed in a number of directions. Some go towards the heat source and die.				
A motile alga swims towards light.				

[5]
[Total: 5]

3 A grebe is a water bird which displays a distinctive courtship ritual.
Male behaviour is distinctive from that of the female.

State **three** advantages to the species of this behaviour. [3]

[Total: 3]

4 The diagrams show stages in the development of a garden pond over a 10-year period.

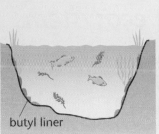  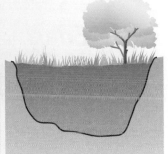

butyl liner		
A hole was dug, lined with butyl liner and new plants were placed in the pond.	Marginal plants grow, spread and die down in the winter. As they rot sediment falls to the bottom of the pond.	After a number of years the pond has completely covered over.
1990	**1995**	**2000**

(a) In 1990 irises, oxygenating pondweed and a water lily were planted in the pond. Algae were not planted but arrived in the pond in some other way.

   (i) What term describes an organism that grows in a new habitat that previously supported no life? [1]

   (ii) After a time the algae produced a thick 'carpet' of growth on the surface of the pond. Explain the effect this may have on organisms under the water. [5]

(b) Describe the stages which took place to produce the stable grassland after 10 years. [4]

[Total: 10]

Chapter 8

Energy and ecosystems

The following topics are covered in this chapter:

- Energy flow through ecosystems
- Nutrient cycles

8.1 Energy flow through ecosystems

After studying this section you should be able to:

- understand the roles of producers, consumers and decomposers in food chains
- understand the flow of energy through an ecosystem

LEARNING SUMMARY

Food chains and energy flow

OCR ▷ 5.3.1

Before energy is available to organisms in an ecosystem, photosynthesis must take place. Sunlight energy enters the ecosystem and some is available for photosynthesis. Not all light energy reaches photosynthetic tissues. Some totally misses plants and may be absorbed or reflected by items such as water, rock or soil. Some light energy which does reach plants may be reflected by the waxy cuticle or even miss chloroplasts completely! The energy that is trapped by photosynthesis and converted into biomass is called the **gross primary productivity** (GPP).

> Around 4% of light entering an ecosystem is actually used in photosynthesis.
>
> **KEY POINT**

The green plant uses the **carbohydrate** as a first stage substance and goes on to make **proteins** and **lipids**. Plants are a rich source of nutrients, available to the herbivores which eat the plants. Some energy is not available to the herbivores because green plants **respire** (releasing energy).

The energy that is available to herbivores is called the **net primary productivity** (NPP). It is calculated as follows:

Net primary productivity = gross primary productivity – energy lost in respiration

Energy is also lost from the food chain as **not all parts** of plants may be **consumed**, e.g. roots.

Food chains and webs

Energy is passed along a food chain. Each food chain always begins with an **autotrophic** organism (producer), then energy is passed to a primary consumer, then a secondary consumer, then a tertiary consumer and so on.

direction of energy flow →

Producer → primary consumer → secondary consumer → tertiary consumer
(herbivore)   (1st carnivore)   (2nd carnivore)

The following example shows three food chains linked to form a food web.

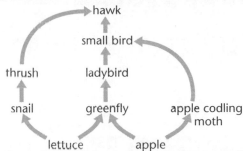

Note that a small bird is a secondary consumer when it eats apple codling moths but a tertiary consumer when it eats greenfly.

Each feeding level along a food chain can also be represented by a **trophic level**. The food chain below is taken from the food web above and illustrates trophic levels. Energy may be used by an organism in a number of different ways:

* respiration releases energy for movement or maintenance of body temperature, etc.

* production of new cells in growth and repair

* production of eggs

* released trapped in excretory products.

The producers always have more energy than the primary consumers, the primary consumers more than the secondary consumers and so on, up the food web. Energy is released by each organism as it respires. Some energy fails to reach the next organism because not all parts may be eaten.

Examiner's tip! Note that the primary consumer is at trophic level 2. It is easy to make a mistake with this concept. Many candidates do!

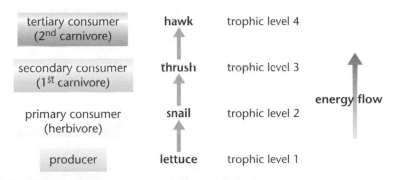

Pyramids of numbers, energy and biomass

A food chain gives limited information about feeding relationships in an area. Actual proportions of organisms in an area give more useful data. Consider a food chain from a wheat field. The pyramid of numbers sometimes does not give a suitable shape. In the example shown below, there are more aphids in the field than wheat plants. This gives the shape shown below (not a pyramid in shape!). A pyramid of biomass is more likely to be a pyramid in shape because it takes into account the size of the organism. It does not always take into account the rate of growth and so only a pyramid of energy is always the correct shape.

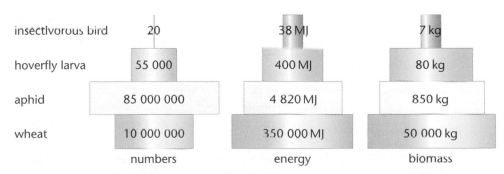

Biomass is the mass of organisms present at each stage of the food chain. The biomass of wheat would include leaves, roots and seeds. (All parts of the plant are included in this measurement.)

The organisms in the above food chain may die rather than be consumed. When this happens, the decomposers use extra-cellular enzymes to break down any organic debris in the environment. Corpses, faeces and parts that are not consumed are all available for decay.

8.2 Nutrient cycles

After studying this section you should be able to:

- *recall how nitrogen is recycled*

The nitrogen cycle

OCR   S.3.1

Nitrogen is found in every amino acid, protein, DNA and RNA. It is an essential element! Most organisms are unable to use atmospheric nitrogen directly so the nitrogen cycle is very important.

There are three parts of the nitrogen cycle which are regularly examined:
- nitrogen fixation in leguminous plants
- nitrification
- denitrification.

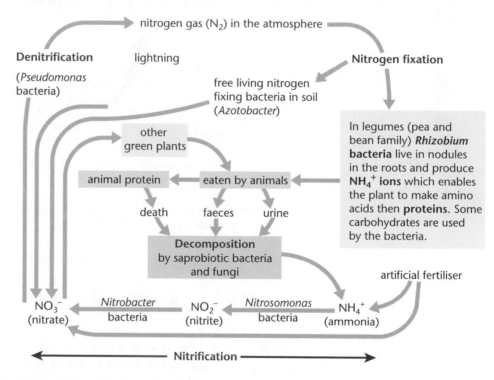

Some important points

The association of *Rhizobium* bacteria with legume plants give advantages to both organisms. This relationship is known as **mutualism**.

- **Nitrogen gas** from the atmosphere is used by *Rhizobium* bacteria. These bacteria, living in nodules of legume plants, convert nitrogen gas into **ammonia** (NH_3) then into amine ($-NH_2$) compounds. The plants transport the amines from the nodules and make amino acids then proteins. *Rhizobium* bacteria gain carbohydrates from the plant, therefore each organism benefits.

Saprobiotic bacteria and fungi secrete extra-cellular enzymes.

- Plants support food webs, throughout which excretion, production of faeces and death take place. These resources are of considerable benefit to the ecosystem, but first **decomposition** by **saprobiotic** bacteria takes place, a waste product of this process is **ammonia**.
- Ammonia is needed by *Nitrosomonas* bacteria for a special type of nutrition (chemo-autotrophic). As a result another waste product, **nitrite** (NO_2) is formed.

The biochemical route from ammonia to nitrate is **nitrification**. This is helped by ploughing which allows air into the soil. Nitrifying bacteria are aerobic. Draining also helps.

- Nitrite is needed by *Nitrobacter* bacteria, again for chemo-autotrophic nutrition. The waste product from this process is **nitrate**, vital for plant growth. Plants absorb large quantities of nitrates via their roots.
- Nitrogen gas is returned to the atmosphere by **denitrifying bacteria** such as *Pseudomonas*. Some nitrate is converted back to nitrogen gas by these bacteria. The cycle is complete!

Sample question and model answer

(a) The sequence below shows how nitrate can be produced from a supply of oak leaves.

$$\underset{\text{leaves}}{\text{dead oak}} \xrightarrow{\underset{\text{decomposers}}{}} \underset{\text{(ammonia)}}{NH_3} \xrightarrow{\substack{\textit{Nitrosomonas}\\ \text{bacteria}}} \underset{\text{(nitrite)}}{NO_2} \xrightarrow{\substack{\textit{Nitrobacter}\\ \text{bacteria}}} \underset{\text{(nitrate)}}{NO_3}$$

(i) Suggest the consequences of death of the *Nitrosomonas* bacteria. [4]

build up of ammonia; build up of dead leaves; death of *Nitrobacter* bacteria; no nitrite/no nitrate

(ii) Name the process by which bacteria produce nitrate from ammonia. [1]

nitrification

(iii) Name **two** populations of organisms not shown in the sequence which would be harmed by a lack of nitrate. [2]

denitrifying bacteria or *Pseudomonas*; plants or producers

(iv) Which organisms fix atmospheric nitrogen on the nodules of bean plants? [1]

(*Rhizobium*) bacteria

(b) (i) A farmer rears pigs by a factory-farming method. Pigs are kept indoors 24 hours per day in warm, confined cubicles.
How can this method result in the production of a greater yield of pork than from animals reared outside? [3]

less energy is released for movement; less energy is used to maintain body temperature; more energy is used for biomass

(ii) Suggest why many consumers object to this factory farming method [1]

cruel or not ethical

Practice examination questions

1 The number of species of grass and the number of leguminous plants growing in two fields was measured over a 10-year period. Field A was given nitrogenous fertiliser each year, but field B was given none. The results are shown in the graphs.

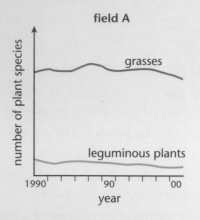

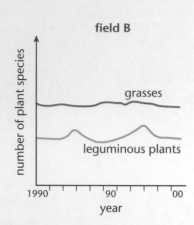

(a) (i) Suggest why there were fewer leguminous plant species in field A. [2]

(ii) Suggest why there were more leguminous plant species in field B. [2]

(b) After the main investigation no fertiliser at all was used in either field. Cattle were allowed to graze in both fields. At the end of five years the number of legume species in each field had decreased. Suggest why the number of legume plants decreased. [1]

[Total: 5]

2 The diagram shows the flow of energy through an ecosystem. The energy units are in kJ.

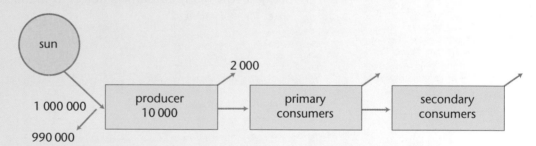

(a) Much of the light energy falling on the producers is not used for photosynthesis. Give two reasons why. [2]

(b) The GPP of the producers is 10 000kJ

(i) What is their NPP? [1]

(ii) Why is there a difference between their GPP and their NPP? [1]

(c) Explain why most food chains are limited to about five trophic levels. [3]

[Total: 7]

Chapter 9

Synoptic assessment

What is synoptic assessment?

You must know the answer to this question if you are to be fully prepared for your A2 examinations!

Synoptic assessment:

- involves the drawing together of knowledge, understanding and skills learned in different parts of the AS/A2 Biology courses
- requires that candidates apply their knowledge of a number of areas of the course to a variety of contexts
- is tested at the end of the A2 course both by assessment of investigative/practical skills and by examinations
- is valued at 20% of the marks of the course total.

In the OCR examinations there will be synoptic question on both Unit 1 and Unit 5 papers.

Practical investigations

You will need to apply knowledge and understanding of the concepts and principles, learned throughout the course, in the **planning, execution, analysis** and **evaluation** of each investigation.

How can I prepare for the synoptic questions?

- **Check out the modules** which will be examined for the synoptic questions.
- Expect **new contexts** which draw together lots of different ideas.
- Get ready to **apply** your knowledge to a new situation; contexts change but the **principles remain the same.**
- In modular courses there is sometimes a tendency for candidates to learn for a module, achieve success, then forget the concepts. Do not allow this to happen! **Transfer concepts** from one lesson to another and from one module to another. Make those connections!
- Improve your powers of analysis – **take a range of different factors into consideration** when making conclusions; synoptic questions often involve both graphical data and comprehension passages.
- Less able candidates make limited conclusions; high ability candidates are able to consider several factors at the same time, then make a **number of sound conclusions** (not guesses!).
- You need to do **regular revision** throughout the course; this keeps the concepts 'hot' in your memory, 'simmering and distilling', ready to be **retrieved** and **applied** in the synoptic contexts.
- The bullet point style of this book will help a lot; back this up by summarising points yourself as you make notes.

Why are synoptic skills examined?

Once studying at a higher level or in employment, having a narrow view, or superficial knowledge of a problem, limits your ability to contribute. Having discrete knowledge is not sufficient. You need to have confidence in applying your skills and knowledge.

Synoptic favourites

The final modules, specified by OCR for synoptic assessment, include targeted synoptic questions. Concepts and principles from earlier modules will be tested together with those of the final modules. You can easily identify these questions, as they will be longer and span wide-ranging ideas.

> **Can we predict what may be regularly examined in synoptic questions?**
>
> 'Yes we can!' Below are the top five concepts. Look out for common processes which permeate through the other modules. An earlier module will include centrally important concepts which are important to your understanding of the rest.

KEY POINT

Check out the synoptic charts

1 Energy release

Both aerobic and anaerobic respiration release energy for many cell processes. Any process which harnesses this energy makes a link.

Examples

Synoptic links

Try this yourself! Think logically. Write down an important biological term such as 'cell division'. Link related words to it in a 'flow diagram' or 'mind map'. The links will become evident and could form the framework of a synoptic question.

- Reabsorption of glucose involves active transport in the proximal tubule of a kidney nephron. If you are given a diagram of tubule cells which show both mitochondria and cell surface membrane with transporter proteins, then this is a cue that active transport will probably be required in your answer.

- Contraction of striated (skeletal) muscle requires energy input. This is another link with energy release by mitochondria and could be integrated into a synoptic question.

- The role of the molecule ATP as an energy carrier and its use in the liberation of energy in a range of cellular activity may be regularly linked into synoptic questions. The liberation of energy by ATP hydrolysis to fund the sodium pump action in the axon of a neurone.

- The maintenance of proton gradients by proton pumps is driven by electron energy. Any process involving a proton pump can be integrated into a synoptic question.

2 Energy capture

Energy: input and output

This has to be a favourite for many synoptic questions. Energy is involved in so many processes that the frequency of examination will be high.

Photosynthesis is responsible for availability of most organic substances entering ecosystems. It is not surprising that examiners may explore knowledge of this process and your ability to apply it to ecological scenarios.

Examples

- Given the data of the interacting species in an ecosystem you may be given a short question about the mechanism of photosynthesis then have to follow the energy transfer routes through food webs.

- Often both photosynthesis and respiration are examined in a synoptic type question. There are similarities in both the thylakoid membranes in chloroplasts and cristae of mitochondria.

- Many graphs in ecologically based questions show the increase in herbivore numbers, followed by a corresponding carnivore increase. Missed off the graph, your knowledge of a photosynthetic flush which stimulates herbivore numbers may be expected.

3 The structure and role of DNA

It is important to know the structure of DNA because it is fundamentally important to the maintenance of life processes and the transfer of characteristics from one generation of a species to the next. DNA links into many environmentally and evolutionally based questions.

- The ultimate source of variation is the mutation of DNA. Questions may involve the mechanism of a mutation in terms of DNA change and be followed by natural selection. This can lead to extinction or the formation of a new species. Clearly there are many potential synoptic variations.

- DNA molecules carry the genetic code by which proteins are produced in cells. This links into the production of important proteins. The structure of a protein into primary, secondary, tertiary and quaternary structure may be tested. All enzymes are proteins, so a range of enzymically based question components can be expected in synoptic questions.

- The human genome project is a high-profile project. The uses of this human gene 'atlas' will lead to many developments in the coming years. The reporting of developments, radiating from the human genome project, could be the basis of many comprehension type questions, spanning diverse areas of Biology. Save newspaper cuttings, search the internet and watch documentaries. Note links with genetic diseases, ethics, drugs, etc.

4 Structure and function of the cell surface membrane

There are a range of different mechanisms by which substances can cross the cell surface membrane. These include diffusion, facilitated diffusion, osmosis, active transport, exocytosis and pinocytosis. Additionally glycoproteins have a cell recognition function and some proteins are enzymic in function. Knowledge of these concepts and processes can be tested in cross-module questions.

- In an ecologically based question the increasing salinity of a rock pool in sunny conditions could be linked to water potential changes in an aquatic plant or animal. Inter-relationships of organisms within a related food web could follow, identifying such a question as synoptic.

- In cystic fibrosis a transmembrane regulator protein is defective. A mutant gene responsible for the condition codes for a protein with a missing amino acid. This can link to the correct functioning of the protein, the mechanism of the mutation and the functioning of the DNA.

5 Transport mechanisms

This theme may unify the following into a synoptic question, transport across membranes, transport mechanisms in animal and plant organs. Additionally, they may be linked to homeostatic processes.

- The route of a substance from production in a cell, through a vessel to the consequences of a tissue which receives the substance, could expand into a synoptic question. Homeostasis and negative feedback could well be linked into these ideas.

Sample questions and model answers

Question 1 *(a short structured question)*

The kangaroo rat (*Dipodomys deserti*) is a small mammal that lives in the Californian desert. It has specialised kidneys so that it can produce a very concentrated urine.

(a) Name the genus that contains the kangaroo rat. [1]

Dipodomys

(b) What is the biological naming system called that gives the kangaroo rat its scientific name? [1]

binomial system

(c) Kangaroo rats have long loops of Henle. In which part of the kidney would you expect to find loops of Henle? [1]

medulla

(d) What is the name of the hormone that controls the concentration of the urine in mammals? [1]

ADH

(e) Which gland releases this hormone into the blood? [1]

pituitary gland

(f) The desert community that contains the kangaroo rat is the final product of succession in California. What is the name of the final, stable community that is produced by succession? [1]

a climax community

[Total: 6]

Question 2 *(a longer, more open-ended question)*

Plants and animals both need to exchange gases with the environment. Describe how animals and plants are adapted for efficient gaseous exchange. [10]

(Quality of written communication assessed in this answer.)

- examples of respiratory surfaces in animals:
  gills/lungs;
  tracheoles in insects;
  surface of protoctists;
  stomata in plants

- large surface area:
  way(s) in which this is achieved e.g.
  many alveoli;
  surface area/volume ratio in protoctists;
  many gill filaments;
  large surface area of leaves;
  many mesophyll cells

- maintenance of diffusion gradients
  way(s) in which this is achieved
  rich blood supply;
  ventilation mechanisms;
  sub-stomatal airspaces;
  spongy mesophyll air spaces;
  use of carbon dioxide in mesophyll cells

- small diffusion pathway
  barriers one cell thick;
  specialised cells, e.g. squamous epithelium;
  thin cell walls of palisade cells

Note, there is one mark available for legible text with accurate spelling, punctuation and grammar.

[Total: 10]

Question 3 *(a longer question of higher mark tariff)*

Different concentrations of maltose were injected in the small intestine of a mouse. The amount of glucose appearing in the blood and the small intestine after 15 minutes were measured. The results are shown in the graph.

Prepare yourself for this type of synoptic question. It cuts across a large part of the specification. Make the links with different ideas. This fact is very important; concepts from AS are needed.

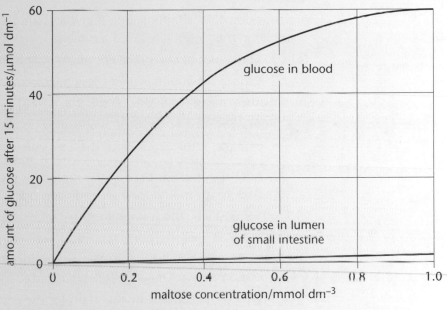

(a) (i) Describe the structure of a maltose molecule [2]

Two molecules of (alpha) glucose;

joined together by a glycosidic bond.

(ii) Maltose is converted into glucose by a hydrolysis reaction. What is a hydrolysis reaction? [1]

A reaction that breaks down a substance by the addition of water.

(b) Describe the effect of different maltose concentrations on the amount of glucose found in the lumen of the small intestine compared to the effect on the amount found in the blood. [2]

Even if you only cover one of these points, you can pick up a second mark by correctly using figures from the graph in your answer.

the maltose concentration has much more effect on the amount of glucose in the blood; the amount of glucose found in the blood is starting to level off but the amount in the lumen is increasing steadily.

Sample questions and model answers (continued)

(c) The enzyme maltase is found on the cell surface membrane of the epithelial cells of the small intestine.

(i) How does the data on the graph indicate that the enzyme is not released into the lumen? [1]

Very little/no increase in the amount of glucose in the lumen.

This is a harder stretch and challenge question.

(ii) Explain why having the enzyme fixed to the cell surface will increase the rate of glucose absorption. [2]

Higher concentration of glucose produced close to intestinal lining;

will increase the concentration gradient between intestine and blood.

[Total: 8]

Question 4

A cow is described as a ruminant. Ruminants are herbivores that have a chamber in their intestines called a rumen.

(a) (i) The rumen of cows contains microorganisms.

Explain the importance of these microorganisms to the cow. [3]

They digest cellulose in the cow's food;

the cow cannot produce the enzyme to digest cellulose;

produce fatty acids that the cow can use.

(ii) After the food has been in the rumen for some time it is regurgitated back to the mouth for a second chewing.

Suggest why this is important. [1]

Increase the surface area for digestion.

(iii) The microorganisms in the rumen produce two waste products, methane and ammonia. The ammonia is converted into urea by the cow's liver.

Why is this conversion important for the cow? [1]

Ammonia is more toxic than urea.

(b) The table shows the amount of methane produced by different domesticated animals

Animal type	Methane production per animal in kg per animal per year	Total methane production in tonnes per year
buffaloes	50	6.2
camels	58	1.0
goats	5	2.4
sheep	6	3.4

(i) Which of the animals in the table are ruminants? Explain how you can tell this. [2]

Buffaloes and camels;

they produce much more methane per animal.

This is a typical synoptic question as it links two different topics, digestion in herbivores and the greenhouse effect!

(ii) Which type of animal in the table is domesticated in the highest numbers? Explain how you worked out your answer.   [2]

Sheep; dividing the total methane production by the production per animal gives the highest number.

(iii) Methane is a potent greenhouse gas.
What is a greenhouse gas?   [2]

A gas that prevents the escape of infra red radiation from the atmosphere; therefore causes the atmosphere to warm.

(iv) It has recently been discovered that methane is released when arctic ice melts.
Explain why people are concerned by this discovery.   [2]

The release of methane would increase global warming;

which in turn would result in the release of even more methane.

[Total: 13]

Practice examination answers

Chapter 1 Energy for life

1

(a) in cytoplasm [1]

(b) pyruvate [1]

(c) 2 ATPs begin the process;
2ATPs are produced from each of the two GP molecules, so −2 + 4 = +2 ATPs net [1]

(d) animal; animal cells produce lactate [1]

(e) oxygen or aerobic [1]

[Total: 5]

2

(a) At this point the amount of carbon dioxide given off by the plant in *respiration*, is totally used by the plant in *photosynthesis*. [2]

(b) compensation point [1]

(c) The continued graph line falls (as light dims); line ends below the horizontal axis (when it's dark!). [2]

[Total: 5]

3

(a) Absorption spectrum is obtained from the amount of each wavelength absorbed by the pigments which made up the chlorophyll of the plant.

Action spectrum is produced by measuring the amount of photosynthesis by the plant for each separate wavelength. [2]

(b) Low amount of photosynthesis because not much light energy absorbed, most is reflected. [1]

(c) Evolution of oxygen, collected by water displacement. [1]

[Total: 4]

4

(a) mitochondrion [1]

(b) NADH [1]

(c) cytochrome [1]

(d) ATP [1]

[Total: 4]

5

(a) (i) rate of photosynthesis is proportional to light intensity; rate limited by amount of light available

(ii) as light intensity increases it results in significantly less increase in the rate of photosynthesis

(iii) rate of photosynthesis has levelled off, no longer limited by light (but other conditions could be limiting!). [3]

(b) Similar shape of graph, begins at origin, but graph line above the given plotted curve. [1]

[Total: 4]

Chapter 2 Response to stimuli

1

(a) (i) IAA (at these lower) concentrations is *proportional* to the angle of curvature of the stem. [1]

(ii) IAA (at these higher) concentrations is *inversely proportional* to the angle of curvature. [1]

(b) *More* IAA causes the cells at side of stem in contact with agar block to elongate more than other side.

So this side grows more strongly bending stem towards the weaker side. [2]

(c) Growth is only stimulated up to a certain high IAA concentration, after this curvature would be inhibited. [2]

[Total: 6]

2

(a) A = actin
B = myosin [2]

(b) action potential reaches sarcomere [1]

(c) both filaments slide alongside each other;
they form cross bridges;
during contraction the filaments slide together to form a shorter sarcomere [2]

[Total: 5]

3

(i) resting potential achieved; [2]
Na^+ / K^+ pump is on

(ii) Na+ / K^+ pump is off;
so Na^+ ions enter axon [2]

(iii) maximum depolarisation achieved; K^+ ions leave [2]

(iv) Na^+ ions leave due to Na^+/ K^+ pump being back on;
this is during the refractory period;

(v) at end of this resting potential re-established;
axon membrane re-polarised [4]

[Total: 10]

Chapter 3 Homeostasis

1 (a)

	Nervous system	Endocrine system
Usually have longer lasting effects		✓
Have cells which secrete transmitter molecules	✓	
Cells communicate by substances in the blood plasma		✓
Use chemicals which bind to receptor sites in cell surface proteins	✓	✓
Involve the use of Na$^+$ and K$^+$ pumps	✓	

[2]

(b) homeostasis [1]

[Total: 3]

2

It increases permeability of; the collecting ducts, and the distal convoluted tubules of the nephron;
- more water drawn out of the collecting ducts;
- by the sodium and chloride ions;
- in medulla of kidney;
- so more water can be reabsorbed back into blood;
- through the capillary network. (max 6) [6]

[Total: 6]

3

(a)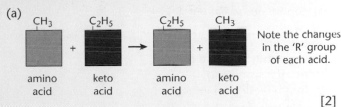

Note the changes in the 'R' group of each acid. [2]

(b) (i) liver [1]

(ii) To make different amino acids with the help of the essential amino acids. [2]

[Total: 5]

4

(a) **B**, because as glucose levels rose after meals they did not decrease enough (this kept the blood glucose level too high) [1]

(b) glucose levels fell after every meal, so glucose must have entered the cells and liver [1]

(c) in the pancreas;
in the β cells of islets of Langerhans (max 2) [2]

[Total: 4]

Chapter 4 Further genetics

1

(a) no immigration and no emigration; no mutations; no natural selection; true random mating, all genotypes must be equally fertile [4]

(b) (i) $q^2 = \dfrac{48}{160}$

$= 0.3$

$q = 0.55$

but $p + q = 1$

so $p = 1 - 0.55$

$= 0.45$

but $p^2 + 2pq + q^2 = 1$

so $0.45^2 + 2 \times 0.45 \times 0.55 + 0.55^2 = 1$

$0.2 + 0.5 + 0.3 = 1$

BB = 0.2   Bb = 0.5 bb = 0.3 [3]

(ii) BB 2000   Bb 5000   bb 3000 [2]

[Total: 9]

2

A (iv), B (iii), C (v), D, (ii), E (i). [Total: 5]

3

(a) triplet [1]

(b) codes for an amino acid, codes for stop or start [2]

[Total: 3]

4

(a) 8 or 4 pairs [1]

(b) (i) During telophase I of meiosis the chromosomes are bivalent/the centromeres are still intact, whereas in telophase II the chromosomes are single [1]

(ii) During telophase of mitosis the chromosomes are in pairs, whereas in telophase II of meiosis they are single (haploid) [1]

(c) the spindle contracts; pulls the centromeres apart; chromosomes begin to be pulled to both poles. [2]

[Total: 5]

Chapter 5 Variation and selection

1

(a) continuous variation; [1]

(b) two from:
genetic;
the nutrition of the mother;
mother's smoking;
mother's alcohol intake;
mother's health; [2]

(c) three from:
heavy babies have higher death rate;
light babies have higher death rate;
so babies of average mass more likely to survive;
they are more likely to have babies of average mass; [3]

(d) Modern techniques can increase survival of light
and heavy babies; they in turn will reproduce; [2]
[Total: 8]

2

(a) **Allopatric speciation** takes place after geographical
isolation;
• the rising of sea level splits a population of animals;
formerly connected by land creating two islands;
• mutations take place so that two groups result in
different species.

Sympatric speciation takes place through genetic
variation;
• in the same geographical area;
• mutation may result in reproductive incompatibility;
• perhaps a structure in birds may lead to a different
song being produced by the new variant;
• this may lead to the new variant being rejected
from the mainstream group;
• breeding may be possible within its own group
of variants. [6]

(b) Mate them both with a similar male, to give them
a chance to produce fertile offspring.
• If they both produce offspring, take a male and
female from the offspring, mate them,
• if they produce fertile offspring then original
females **are** from the same species. [2]
[Total: 8]

Chapter 6 Biotechnology and genes

1

(a) Steam sterilisation;
microorganisms cannot enter through air filter;
nutrients are pre-sterilised before entry into
fermenter. [3]

(b) Contaminant microorganisms enter the fermenter;
compete with the *Penicillium*; fungus;
penicillin yield reduced. [3]
[Total: 6]

2

(a) Identify the specific section of DNA which contains the
gene; this can be done using reverse transcriptase; insert
DNA into a vector/insert into *Agrobacterium tumefaciens*;
this bacterium/this vector then passes the DNA into the
recipient cell. [5]

(b) herbicide kills weeds; which reduces competition; for
light or water or minerals; soya plants unharmed [3]
[Total: 8]

3

(a) Beginning of fermentation process shown.
The microorganisms took time to reach
maximum production but kept at this level.
Nutrients constantly added. [1]

(b) continuous
product amount reaches a constant level;
nutrients at constant level. [2]
[Total: 3]

Chapter 7 Ecology and populations

1

(a) no significant migration;
no significant births or deaths;
marking does not have an adverse effect. [3]

(b) S = total number of individuals in the total population
S_1 = number captured in sample one, marked and released, i.e. 16
S_2 = total number captured in sample two, i.e. 12
S_3 = total marked individuals captured in sample two, i.e. 5

$$\frac{S}{S_1} = \frac{S_2}{S_3} \quad \text{so, } S = \frac{S_1 \times S_2}{S_3}$$

$S = \dfrac{16 \times 12}{5}$ Estimated no. of shrews is 38 [2]

(c) Not very reliable because the numbers are quite low. High population numbers are more reliable. [1]

[Total: 6]

2

	Type of behaviour			
	simple reflex	kinesis	positive taxis	negative luxis
A bolus of food reaches the top of our oesophagus and is swallowed.	✓			
Insects move from a cold dry area to a warm humid one.			✓	✓
Springtails (insects) are subjected to increasingly hot conditions, and react by increasing speed in a number of directions. Some go towards the heat source and die.		✓		
A motile alga swims towards light.			✓	

[Total: 5]

3

The opposite sexes recognise each other;
the grebes will only mate with other grebes so are more likely to produce fertile offspring;
mating is synchronised, to coincide with ovulation. [3]

[Total: 3]

4

(a) (i) pioneer or primary coloniser [1]

(ii)
  • algae cut off light from plants underneath;
  • they die as a result;
  • bacteria or fungi or saprobiotics decay them;
  • they use a lot of oxygen;
  • fish die due to not enough oxygen;
  • blood worms increase in number as they are adapted to small amounts of oxygen.

[any 5 points] [5]

(b) • marginal plants or irises were introduced;
  • they spread;
  • each year the foliage died and rotted;
  • this organic material or humus added to the soil or mud;
  • secondary colonisers spread from other areas;
  • succession took place.

[any 4 points] [4]

[Total: 10]

Chapter 8 Energy and ecosystems

1

(a) (i) When given fertiliser the grasses competed for resources better that the legumes; some legume species could not grow in these conditions. [2]

(ii) Without fertiliser the grass species did not have enough minerals so did not compete as well; the legumes fixed nitrogen in root nodules so could grow effectively. [2]

(b) Cows grazed on some species more than others/ perhaps trampling by cattle destroyed some species but others were tougher and survived/perhaps waste encouraged the growth of some species whereas others were destroyed. [1]

[Total: 5]

2

(a) Reflected from leaf; passes through leaf; wrong wavelength. [2]

(b) (i) NPP = 8000kJ
(ii) NPP is GPP minus losses from respiration. [2]

(c) Energy is lost at each transfer; Through excretion / egestion / uneaten parts; Not enough energy left for further levels [3]

Notes

Notes

Notes

Index